Albert ALOMBA A.

INTRODUCTION TO GEOLOGY

Albert ALOMBA A.

INTRODUCTION TO GEOLOGY

For students of the scientific humanities

ScienciaScripts

Imprint
Any brand names and product names mentioned in this book are subject to trademark, brand or patent protection and are trademarks or registered trademarks of their respective holders. The use of brand names, product names, common names, trade names, product descriptions etc. even without a particular marking in this work is in no way to be construed to mean that such names may be regarded as unrestricted in respect of trademark and brand protection legislation and could thus be used by anyone.

Cover image: www.ingimage.com

This book is a translation from the original published under ISBN 978-620-6-72423-0.

Publisher:
Sciencia Scripts
is a trademark of
Dodo Books Indian Ocean Ltd. and OmniScriptum S.R.L publishing group

120 High Road, East Finchley, London, N2 9ED, United Kingdom
Str. Armeneasca 28/1, office 1, Chisinau MD-2012, Republic of Moldova, Europe
Printed at: see last page
ISBN: 978-620-8-23775-2

Contents

BY THE SAME AUTHOR

Cell biology for the 3eme year of scientific humanities
Biology variability approach for the students of senior six advanced level
Chemistry for the 1ere year of scientific humanities
Chemistry for the 3eme year of scientific humanities
Chemistry for the 4eme year of scientific humanities
Ecology for science humanities students
Evolution of living beings for the 4eme year of scientific humanities
Introduction to taxonomy for science humanities students
Population and natural resources for senior six advanced level
Review of plant physiology for students of senior six advanced level
SVT for the 3eme scientific humanities year
SVT for the 4eme scientific humanities year
SVT for grade 8 basic education
SVT for year 7eme basic education
Practical biology exercises for science humanities students

FOREWORD

This book, entitled *Introduction to Geology*, is intended for humanities students in the scientific section in the Democratic Republic of Congo.
The book covers a wide range of topics, including the Earth in the solar system, earthquakes, volcanoes, the geological resources of the Democratic Republic of the Congo, soils, rocks, geological scales and a little human paleontology.
The book covers all the chapters of geology, from the first year of science to the fourth year of science. And it does so in line with the essential knowledge set out in the new Life and Earth Sciences syllabus currently in force in our country.
Given that the change in the biology syllabus to Life and Earth Sciences is not quickly accompanied and supported by new adapted textbooks, the publication of this book will enable students in the scientific section to prepare well for the State Examinations and SVT teachers will be able to use it to prepare their lessons.
We think the students in the science section will make good use of it. We welcome your opinions and suggestions.
Albert ALOMBA A.

We dedicate this book to you, Papa **Paul OTEPA w' OMEDJELA**.

INTRODUCTION

Geology is the science of the Earth. It is particularly interested in the earth's crust through its components, its history and its evolution.

It covers a number of fields, the main ones being *mineralogy* (the study of minerals), *paleontology* (the study of fossils), *sedimentology* (the study of sedimentary deposits), *petrology* (the study of rocks), *geomorphology* (the study of landforms), *geochemistry* (the study of rock chemistry) and *tectonics* (the study of crustal deformation).

It also includes specific fields that are closely related to the above, such as *volcanology* and *seismology*, natural and man-made environmental risks (soil and groundwater pollution, etc.), and *hydrogeology*.

Related definitions: *cosmology* (study of the origin and evolution of the universe), *astronomy* (study of the relative positions, movements, structure and evolution of the stars) and *astrology* (set of beliefs and practices based on the interpretation of the position and movements of the stars, not a science but metaphysical practices).

CHAPTER 1

THE EARTH

1.1 THE EARTH IS A PLANET IN THE SOLAR SYSTEM

The Earth is a planet in the solar system. The sun is a star around which various objects revolve (planets and their satellites, dwarf planets, asteroids, comets). They vary in size, chemical composition and internal activity. Some planets have gaseous or liquid envelopes.

A *planet* is a celestial body of spherical shape, not luminous in itself, which orbits the sun (or a star) and has eliminated any body likely to move in a close orbit. An *exoplanet* is located outside the solar system.

Figure 1.1 *The solar system*

The planets in our solar system are divided into two families: telluric planets and gaseous planets.

The *telluric planets* are : Mercury, Venus, Earth and Mars. They have a solid rocky surface and are essentially made up of silicates and iron.

The *gaseous planets,* also known as giant planets because of their larger size compared to the terrestrial planets, are : Jupiter, Saturn, Uranus and Neptune. They are essentially made up of hydrogen and helium.

A new category of planets, called *dwarf planets*, groups together a wide variety of objects in our universe or solar system. There are several known dwarf planets, such as Ceres (in the asteroid belt between Mars and Jupiter) and Pluto, Eris, Haumea, Makemake ... (in the Kuiper belt), although their total number is not exactly known.

Pluto, formerly classified as a planet, has been considered a dwarf planet since August 2006.

Figure 1.2 *The asteroid belts*

Asteroids are similar in composition to terrestrial planets, but smaller and more irregular in shape. Several thousand have been recorded.

Most evolve within a *belt* between the orbits of Mars and Jupiter, while others, further out, form the *Kuiper belt*, sometimes called the Edgeworth-Kuiper belt, an area of the Solar System extending beyond the orbit of Neptune, between 30 and 55 astronomical units (AU).

The Kuiper belt should not be confused with the *Oort cloud*, a zone that is still theoretical and thought to be a thousand times further away. The objects in the Kuiper belt, along with the scattered objects and any potential members of the **Hills and Oort clouds**, are collectively known as Trans-Neptunian objects.

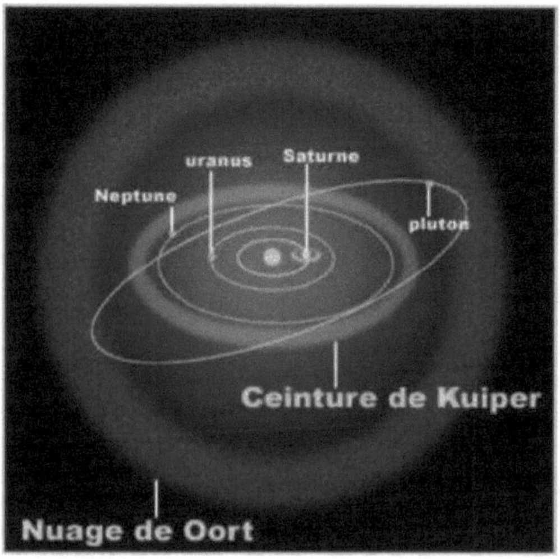

Figure 1.3 *The Oort cloud*

Comets are clusters of ice (frozen water and gas) and dust. They are located much further from the Sun than planets, so they can deviate from their trajectory and approach the Sun. In this case, some of the ice vaporises and the gas and dust form the tail.

The *Earth* has a solid surface. This has enabled it to retain traces of the meteorite impacts that have struck it since its formation. The rocky nature of its surface is underlined by the presence of volcanoes, mountain ranges and sustained internal geological activity. These characteristics are shared by all the telluric planets.

Here are a few parameters for the Earth: average distance from the Sun: 149.6 million km (= 1 AU), equatorial diameter: 12,757 km. Density: 5.52. Obliquity of the axis of rotation in relation to the ecliptic: 23.4°, ground temperature: from -90° to +60°C and average composition: silicates and iron.

The *moon* - our moon - is the Earth's natural satellite. It is located 384,400 km from the Earth and has a diameter of 1,738 km, or a quarter of the Earth. The moon's surface, as seen from Earth, has two distinct aspects: dark areas called *seas* and lighter areas called *land* or *continents*. The lands or continents are the parts of the Moon most heavily impacted by meteorites; observed for the first time by *Galileo* in 1610, the rocks have been dated back to around 4.4 Ga. The seas correspond to rocky surfaces of a basaltic nature that are slightly more recent and less bombarded by meteorites; from 3 to 4 Ga.

In general, 220 natural satellites called *moons* have been identified. Saturn has 82 natural satellites (Titan with its dense atmosphere, winds, rain and rivers of methane), ahead of Jupiter, which has 79 (Io, Europa,

Ganymede and Callisto, etc.). Uranus has 27 moons (Titania , Oberon, Ariel, Umbriel, Miranda ...), Neptune has 14 (Triton, Nereide, Larissa, Hippocampe, .), Mars has 2 (Phobos and Deimos) and the Earth has one. Ours is simply called the Moon. Mercury and Venus are the only planets in the solar system without a moon.

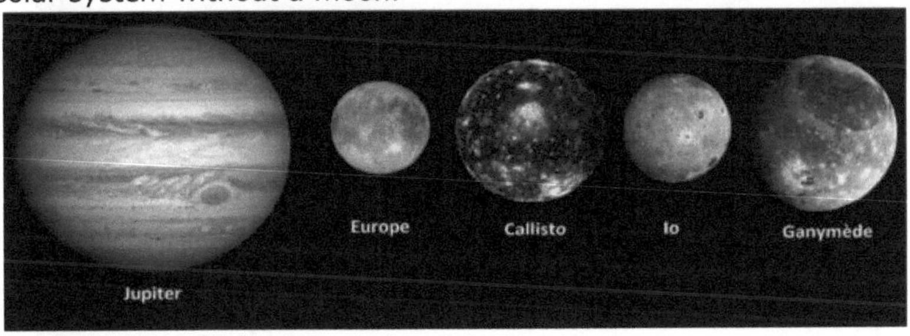

Figure 1.4 *Jupiter's moons*

The *Plutonian system*, a component of the outer solar system. It comprises : Pluto and Sharon, its main moon, and 4 other much smaller satellites: Styx, Nix, Kerberons and Hydra.

A *constellation* is a group of stars that are apparently close together from the Earth, and a conventionally delineated region of the sky that includes this group of stars.

The 12 constellations of the zodiac (named by humans from the earth) are: Aries, Taurus, Gemini, Cancer, Leo, Aquarius, Libra, Scorpio, Sagittarius, Capricorn and Pisces.

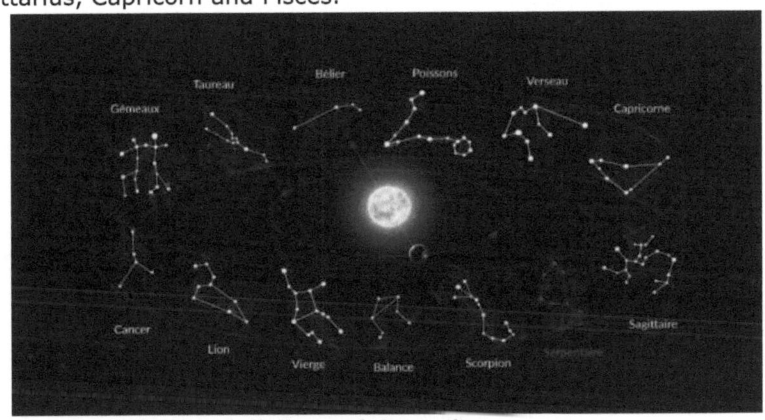

Figure 1.5 *All the constellations of the zodiac*

Galaxies are sprawling systems made up of dust and millions of stars, all held together by gravity.

The Earth is located in the Milky Way galaxy, which contains between 200 and 400 billion stars, including the Sun, and, according to some sources, more than 100 billion planets. The Milky Way has been observed with the naked eye since the dawn of time.

The Milky Way, the Large Magellanic Cloud, the Small Magellanic Cloud and the Andromeda galaxy are visible with the naked eye.

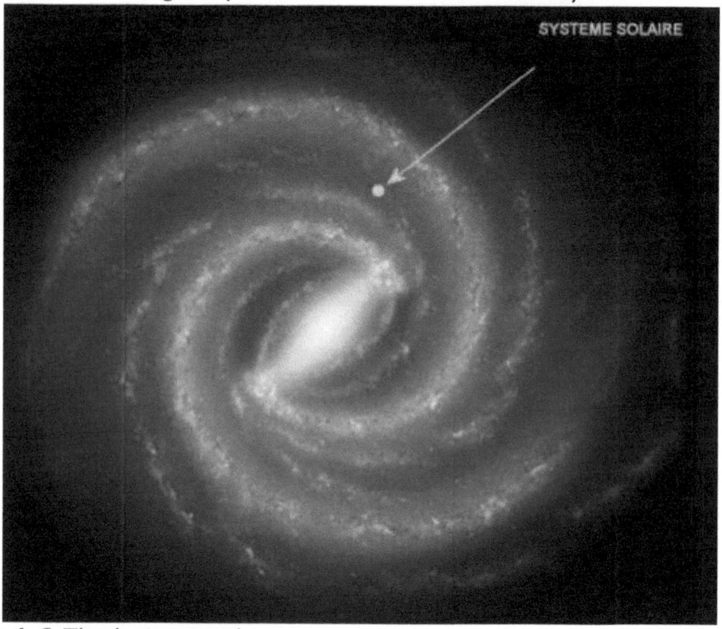

Figure 1.6 *The lactee track*

Table 1.1 *Planets in the solar system*

Planete	Mercure	Venus	Earth	March	Jupiter	Saturn	Uranus	Neptune	Pluto
Distance In the sun : - in millions of km -in U.A. (astronomical units)	from 46 a 69,8 of 0.307 a 0,466	108,21 0,723	149,6 1	227,94 1,524	778,34 5,203	1 426 9,555	2869,6 19,22	4 496 30,1	of 4,431 a 7 369 from 29.55 a 49,25
Average radius (km)	2 440	6 052	6 371	3 390	69 911	58 232	25 362	24 622	1 195
Weight x10^{24} kg	0,330	4,868	5,9736	0,642	1 898	568,5	86,83	102,43	0,012
Mass /M_T	0,055	0,815	1	0,107	318	95,2	14,5	17,1	0,002
Sideral day length	58,65j	243,02j	23h56min	24,623h	9,925h	10,656h	17,240h	16,110h	6,4j
Dureede revolution around the sun	87,969j	224,701j	365,256j Or 1 year	686,980d Or 1.88 years	4,332.59j Or 11.86 years	10 759.2j Or 29.46 years old	30,688.5d Or 84 years old	60 182.3d Or 164.8 years	90,469.7d Or 248.5 years

1.2 STRUCTURE OF THE EARTH

Overland cruise from 0 й 150 km

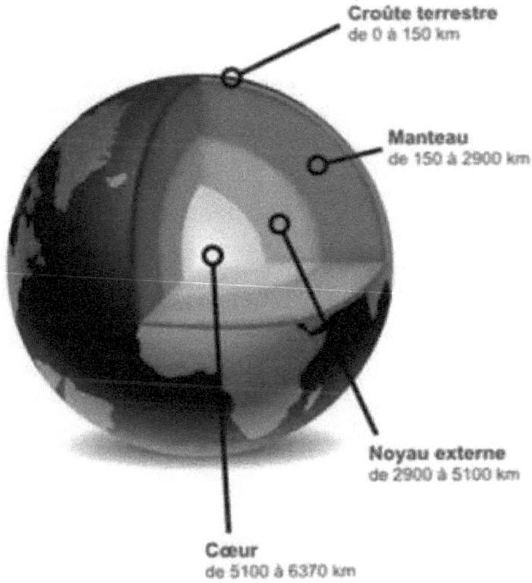

Figure 1.7 *The structure of the Earth*

1.2.1 THE INTERNAL STRUCTURE OF THE GLOBE

The earth is made up of different concentric envelopes with different compositions and different rheological behaviours (rheology: the study of the deformation of bodies).

a. A gaseous envelope, the *atmosphere*.

In the first fifty kilometres above the ground, the Earth's atmosphere is structured in two layers. The troposphere extends from the surface of the ground to the tropopause at an altitude of around ten kilometres, where the temperature decreases with altitude (down to around -55°C towards the tropopause).

The troposphere is relatively low in ozone (the maximum ozone content is never more than 1/1,000th of the CO_2 content) and contains water vapour in variable proportions (a few %). Large convective cells animate the air masses.

This is followed by the stratosphere, from the tropopause to around fifty kilometres above sea level, where the mesosphere begins. The base of the mesosphere is swept by strong tangential air currents known as jets, and it absorbs very little solar radiation, so the temperature drops sharply.

Above an altitude of 80 km, we find the thermosphere, this time with a sharp rise in temperature.

Much further away, at around 500 km altitude, the sun's rays decompose the molecules: this is the ionosphere.

b. A discontinuous liquid envelope, the *hydrosphere*.

c. The *earth's crust*, which can be continental or oceanic.

The *continental crust* has a density of 2.7 in its upper part and 2.8 in its lower part, and its average composition is close to that of granite. It is around thirty km thick, and can be as much as 60 to 70 km thick in orographic zones (mountains).
The *oceanic crust* has a density of around 3, is basaltic in composition and is around 7 km thick.
The continental crust forms a series of blocks that float on the (denser) *mantle* by means of *a hydrostatic-type* mechanism. Any relief in the crust is therefore composed of a root sinking into the mantle, like an iceberg in the sea. This hydrostatic equilibrium is called *isostasy*.
The *oceanic crust* also rests on the mantle and is also in isostatic equilibrium with it.
The differences in thickness and density between the continental crust and the oceanic crust mean that the surface of the latter is usually about 5,000 metres lower on average than that of the continents.
d. The *mantle*, of *peridotitic* composition (a rock composed mainly *of olivine*), comprises :
- A *solid* upper part (density 3.4). The crust and the upper solid part of the mantle make up the *lithosphere*, which is about 100 km thick under the oceans and 150 km thick under the continents.
The *Mohorovicic discontinuity, or Moho,* separates the crust from the lithospheric part of the mantle.
A *discontinuity* is a boundary between two rock formations with distinctly different properties. The lithosphere is divided into *plates*. It can deform plastically at depth.
- A lower, non-solid (but non-liquid) part, the *asthenosphere*, which is able to flow at depth, up to around 2,900 km.
The speed of P waves exceeds 13km/s, at which point the *Gutenberg discontinuity* marks the boundary between the mantle and the outer core: it is both a chemical and a physical discontinuity (the lower mantle is solid, while the core behaves like a fluid, since it does not allow S waves to propagate).
The increase in mantle *density* at depth corresponds to a change in the crystalline (and therefore mineralogical) composition of the peridotite compounds that make it up, under the action of pressure.

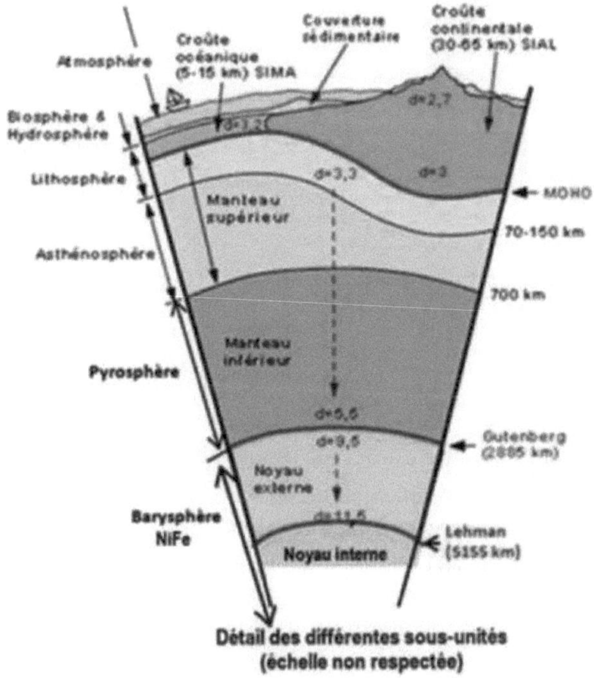

Figure 1.8 *Earth layers*

e. The *outer core, which is liquid and* mainly composed of iron, is 2,900 to 5,100 km long and has a density of 9.8 to 12.

The *Lehmann discontinuity* separates the inner core from the outer core: these two layers of the globe have almost the same chemical composition but different physical properties (the inner core is solid whereas the outer core is fluid).

f. The *inner core*, or *grain*, is *solid*, made of iron and nickel, and extends 5,100 km into the centre of the Earth, with an average density slightly greater than 12.

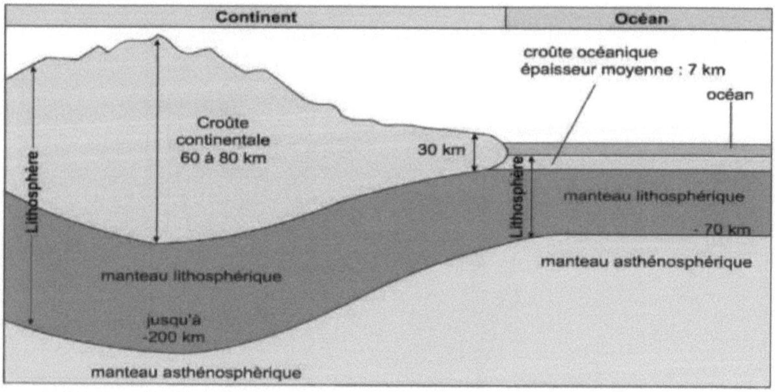

Figure 1.9 *The Earth's crust*

1.2.2 THE COMPOSITION OF THE CRUST AND THE CORE

Thousands of chemical analyses of oceanic and continental crusts have been carried out, providing average compositions.

Analyses of the mantle are based essentially on the few zones of effluvium in the lithospheric part of the mantle and on rock fragments (enclaves) embedded in magmas that originate in the asthenosphere and rise to the surface.

Rocks are the constituent materials of the lithosphere and the vast majority of them are composed of an assemblage of minerals. The minerals that make up rocks are essentially *silicates* and *aluminosilicates*, with *carbonates* being a minor component. Other minerals are only incidental or occasional constituents.

Silicates and aluminosilicates

Quartz: SiO_2

Feldspars: $KAlSi_3O_8$ (orthoclase), $NaAlSi_3O_8$, $CaAl_2Si_2O_8$ (plagioclase)

Micas (white and black): K, Mg and Fe aluminosilicates

Pyroxenes: silicates and aluminosilicates of Ca, Mn, Na, K, Mg, Al, Fe

Amphiboles: same as pyroxenes + OH groups$^-$

Olivines: silicates of Fe, Mg, Mn

Chlorites: same as black micas $+OH^-$

Clay minerals: aluminosilicate hydrates where Al can be replaced by Mg and Fe, sometimes with Na and K.

Carbonates

Calcite: $CaCO_3$

Dolomite: $CaMg(CO_3)_2$

1.2.3 ROCK CLASSIFICATION

Rocks are *classified* on the basis of descriptive characteristics (mineralogical composition, structure) and genetic characteristics (how they are formed).

The most common *rocks* are divided into :

- Rocks of *sedimentary* origin,
- Rocks of *magmatic* origin,
- Rocks of *metamorphic* origin: sedimentary and magmatic rocks that have been transformed by the action of temperature and pressure.

Table 1.2 *The main minereaux*

Minéraux	Formes dans les roches	Couleur éclat	Lumière polarisée non analysée	Lumière polarisée analysée	Formule chimique
Quartz	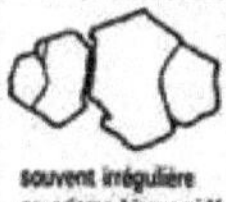 souvent irrégulière ou prisme bipyramidé	incolore, gris limpide éclat gras	incolore limpide	gris clair	SiO_2
Feldspath potassique (orthose)	 mâcle (1) de Carlsbad	blanc, blanc jaunâtre ou rosé aspect moiré	incolore plus ou moins trouble	blanc, gris ou noir	Si_3AlO_8K
Plagioclases Feldspaths calco-sodiques	 mâcle multiple	blanc, gris clair	incolore à gris	blanc, gris ou noir	albite (Si_3AlO_8) Na anorthite $(Si_2Al_2O_8)$ Ca entre les deux tous les intermédiaires existent
Feldspathoïdes	 néphéline  fines stries leucite formes hexagonale ou cubique	incolore à grisâtre éclat vitreux	incolore	gris foncé	néphéline $(SiAlO_4)_4$ Na_3K leucite (Si_2AlO_6) K
Biotite (mica noir)	lamelles hexagonales ou bâtonnets 	brun foncé mordoré	brun foncé jaune clair	bleu, vert, jaune, rouge, atténué par sa couleur propre	$[(Si_3AlO_{10}]$ $(FeMg)_3$ $K(OH)_2$
Muscovite (mica blanc)	clivages (2) fins et nets	incolore	incolore	bleu, vert, jaune teintes vives	$[(Si_3AlO_{10}]$ Al_2 $K(OH)_2$
Pyroxène (augite)	 clivages à 87°	brun noir ou brun verdâtre	jaune pâle	jaune-orangé-rouge, atténué par la couleur propre	$(SiAlO_3)_2$ Ca (FeMgAl)
Amphibole (hornblende)	 clivage à 124°	teinte sombre verdâtre éclat vitreux	brun foncé vert jaunâtre	bleu, vert, jaune, rouge, atténué par la couleur propre	$(Si_6Al_2O_{22})$ $[(MgFe)_4$ $AlCa_2]$ Na (OH_2)
Péridot (olivine)	 plages craquelées	vert olive éclat vitreux	incolore limpide	teintes vives bleu, vert, jaune, rouge	SiO_4 $(MgFe)_2$

1. **Mâcle** : groupement de deux ou plusieurs cristaux de même espèce minérale orientés différemment l'un par rapport à l'autre.
2. **Clivage** : aptitude d'un minéral à se débiter en suivant des plans particuliers appelés plans de clivage. Ces plans sont souvent parallèles à des faces du cristal.

1. Mld*; a grouping of two or more crystals of the same mineral spacing which differ from one another.

[2]- Qhrage: the ability of a nindral 4 to be delimited by particular planes called the ctvage plane*. These planes* often appear as faces of the cnsaf

1.3 THE AGE OF THE EARTH

Astrophysicists *believe* that the history of our universe began *15 million years* ago, in a very hot and very dense environment, with a great explosion, a *great boom*, known as the *big bang*.

The Big Bang cosmological model states that the Universe was denser and hotter in the past and is now expanding. This theory first surfaced in the office of the Russian physicist and mathematician *Alexander Friedmann*, in 1922.

Alexander Friedmann
(1888 - 1925)

The Big Bang theory attempts to explain the first instants of the Universe. It was first put forward, in its current form*, by Abbé Georges Lemaitre*. It refers to the beginning of the expansion of the Universe, around 13.8 billion years ago, when the Universe was extremely dense and hot.

Georges Lemaitre
(1894- 1966)

The resulting rapid drop in temperature led to the formation of the first chemical elements, essentially hydrogen and a little helium. This primordial matter agglomerated into gigantic clusters, the galaxies of the future.

During the 7 to 8 billion years preceding the appearance of our solar system, a multitude of stars produced all the atoms and chemical elements present in the universe, each time dispersing colossal quantities of gas and dust in an explosion.

Around 4.55 billion years ago, in our galaxy, the Lactean Way, such a super nova explosion undoubtedly led to the formation of a large cloud, the pre-solar nebula, which condensed into a hotter ball: the sun star.

Under the effect of solar radiation, the volatile elements of the *nebula* (cosmic cloud: hydrogen and helium) are forced towards the periphery of the solar system. The giant planets at the edges of the solar system (Jupiter, Saturn, etc.) have trapped these gases.

The Earth acquired its present size 4.44 billion years ago. The reconstruction of this history has been made possible by the study of meteorites (4.4 billion years old). Their mineralogical composition corresponds to an assemblage of silicates and iron, the essential constituents of the Earth.

For astrophysicists, therefore, meteorites represent witnesses, objects that clumped together in the solar nebula during the formation of the solar system.

REVISION QUESTIONS ON CHAPTER 1

1. Define the following sciences: geology, mineralogy, sedimentology, petrology, geomorphology, geochemistry, tectonics and hydrology, astronomy, astrology.
2. Explain these concepts: solar system, planet, terrestrial planet, gas planet, dwarf planet, asteroid, Kuiper belt, astronomical unit, year of light, Oort cloud, comet, moon, constellation, zodiac constellations, galaxy and exoplanet.
3. Name the planets in our solar system and their main moons. Give the characteristics of our moon, the moon Europa and the moon Titan. Locate the following celestial bodies in our solar system: Ceres. Eris and Pluto.
4. Describe the inner shells of the globe.
5. What is a discontinuity? Locate the two main discontinuities on Earth.
6. Explain the mechanism of isostatic realisation or simply isostasy.
7. Use a diagram to describe the parts of the Earth's crust.
8. What is a rock? Name the 3 types of mineral that make up a rock.
9. Name 10 minerals that can be used to make rocks in the Earth's crust.
10. What is the approximate age of the Earth? The approximate age of our solar system? Explain the big-bang theory.

CHAPTER 2

THE SEISM

2.1 SEISMIC WAVES

2.1.1 The origin of waves

The origin of earthquakes is almost always the same: a break in the rock at a point called a *focus*, which is usually located within the first 100 km of the earth's outer layer.

This rupture occurs *on a fault*, in an area where the earth's crust is subject to *tectonic stress* (slow movement of two rigid blocks relative to each other).

Forces build up; slowly the rocks deform and break, and the frictional movements of the faults suddenly release energy, energy that is dissipated both in the form of *heat* and in the form of *seismic waves* that propagate in all directions from the focus.

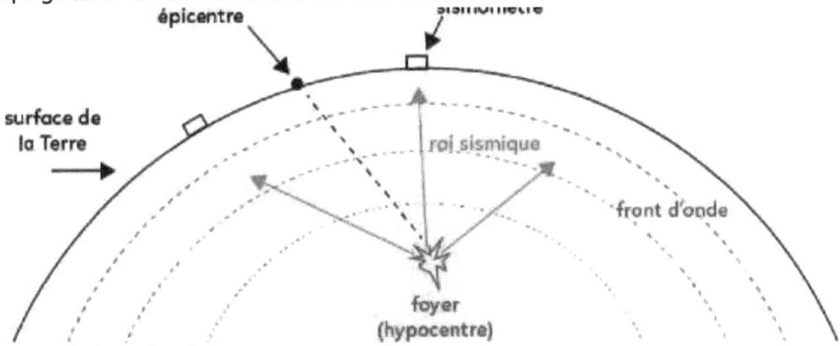

Figure 2.1 *Seismic wave propagation*

2.1.2 Richter scale

Historically, *the Richter scale* was one of the first attempts to numerically assess the intensity of earthquakes, thanks to the Richter magnitude, which measures the seismic energy radiated during an earthquake.

The *Richter scale* is a unit for measuring the force of an earthquake. It was established in 1935 by the American geologist Charles Francis Richter.

In principle, the Richter scale has no upper limit, but geologists believe that the maximum resistance of the earth's rocks prevents any earthquake exceeding 10.

Degree 1 corresponds to the weakest tremors, barely felt, while degree 12 expresses total destruction of the landscape.

Table 2.1 *Richter scale*

Magnitude	Effets engendrés
9	Destruction totale à l'épicentre, et possible sur plusieurs milliers de km
8	Dégâts majeurs à l'épicentre, et sur plusieurs centaines de km
7	Importants dégâts à l'épicentre, secousse ressentie à plusieurs centaines de km
6	Dégâts à l'épicentre dont l'ampleur dépend de la qualité des constructions
5	Tremblement fortement ressenti, dommages mineurs près de l'épicentre
4	Secousse sensible, mais pas de dégâts
3	Seuil à partir duquel la secousse devient sensible pour la plupart des gens
2	Secousse ressentie uniquement par des gens au repos
1	Secousse imperceptible

Charles Francis Richter (1900-1985)

2.1.3 Seismographs

Most *seismographs*, or recording devices, are of the *pendulum* type, i.e. made up of a rigid mass suspended from a wire or a spring, depending on whether you want to record tremors with a major horizontal or vertical component.

When a jolt shakes the support, and therefore the recording cylinder, the mass tends to remain immobile by virtue of its inertia.

Current seismographs amplify ground movements by around 10,000 for long-period oscillations (20 s) and 400,000 for short-period oscillations.

The ground movements caused by the arrival of seismic waves can only be known, in magnitude and direction, if they are related to a system of three orthogonal coordinates. Three seismographs must therefore be used:

- A horizontal seismograph oriented north-south,
- A horizontal seismograph oriented east-west,
- A vertical seismograph.

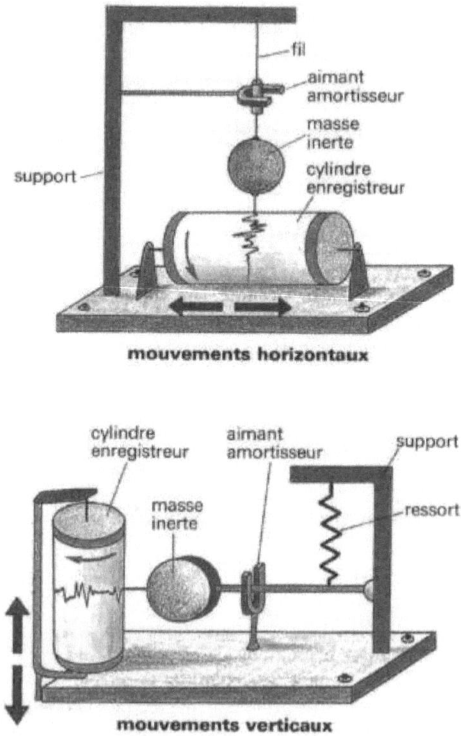

Figure 2.2 *Mechanical seismographs*

2.1.4 Types of seismic waves

The shaking of the earth's crust by an earthquake is reflected at the recording site by a *trace*, the *complexity of* which is due to the successive inscriptions of waves of *different natures* which have travelled across the globe, followed *different trajectories* and crossed, at *different speeds*, zones with *different physical properties*.

While the initial shock lasts a few *seconds*, the arrival of the resulting waves can *last several minutes* at seismographs several thousand kilometres away from the *epicentre*, the point on the earth's surface located vertically above the *focus* or *hypocentre*, where the shaking was at its strongest.

The study of the propagation of refracted seismic waves has made a major contribution to our knowledge of the *internal structure* of the Earth. By analysing *the trajectories* of seismic lines and the propagation *velocities* of seismic waves, a number of *discontinuities* within the Earth's interior can be identified.

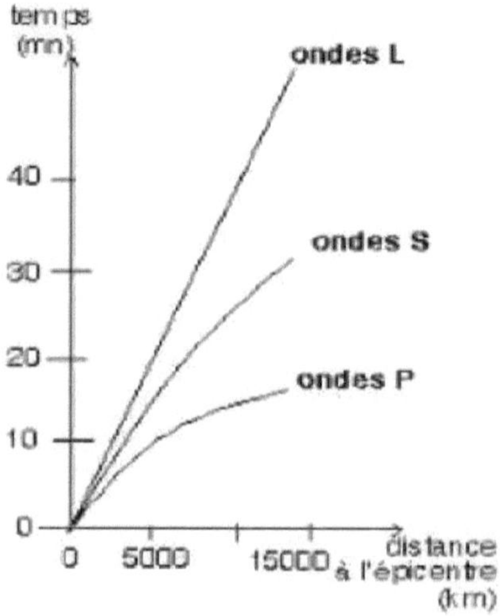

Figure 2.3 *The hodochron curves are independent of earthquake location.*

When the waves are recorded far enough away from the epicentre, *three types of wave* can be easily identified.

A. P waves

P waves or *primary waves* are the fastest waves, the first to register on the seismograph. They are *compressional* waves capable of propagating through both *solids* and *fluids,* including *the atmosphere*. P waves are responsible for the muffled rumbling that can be heard at the start of an earthquake.

B. S waves

S waves, or *secondary waves*, are the second group of waves and are transverse to the direction of propagation. They are only transmitted by *solids*, as liquids offer no resistance to shear.

C. L waves

L waves, or *long waves*, form a third train. They are distinguished from the other two (P and S) by their *large amplitude*. L waves travel only in the *surface layer of* the globe.

2.2 THE MOBILITY OF THE EARTH'S CRUST

The idea of the vertical mobility of the continental crust has been accepted since the last century; that of the horizontal mobility of the containers was formulated at the beginning of the century by *Alfred Wegener*.

This last theory, long opposed by geologists the world over, was finally irrefutably proven in the late 50s by *paleomagnetism* (the study of recordings of the Earth's past magnetic field).

2.2.1 Vertical mobility

A. Isostasy

The most spectacular example of *vertical mobility of* the continental crust can be seen in *Scandinavia*:

- Historians tell us that in one century, the *Gulf of Bothnia* rose by 9 mm per year;
- Geologists are finding that 12,000-year-old beaches are now 400 m above current sea level.

Geologists explain these phenomena by the differences in density and consistency between the continental crust and the underlying layers of the globe.

Figure 2.4 *The high coast of the Gulf of Bothnia*
(Source: Travel and Nature on the Overblog portal).

Over the last million years, a large ice cap (representing a considerable mass) covered Scandinavia.

The lithosphere, part of which is the crust (density 2.7), has been embedded in the asthenosphere, which has a certain fluidity and a density of 3.3.

The melting of the cap began 12,000 years ago and was, on a geological scale, very rapid. Like a cork being pushed into highly viscous oil. Scandinavia then began to rise, a process which is now almost complete. This *rebalancing* according to *density* is called *isostatic readjustment*.

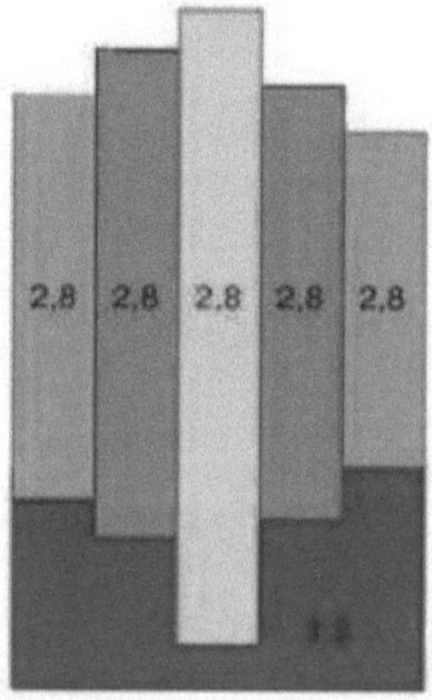

Figure 2.5 *Isostasy modelling*

B. Subsidence

The *coalfield of* northern France is a remarkable example of subsidence. Over a depth of 3,500 m, sandstone and shale formations with interbedded coal veins have accumulated.

More detailed observation shows that, at certain levels, freshwater fossil shales and sandstones have been replaced by marine fossil layers, which are not very thick (1 m maximum) but very extensive.

In addition, at the base of each layer of coal, the remains of tree trunks from the coalfield appear to be in place, with their roots sinking into the sterile layer below.

How are we to interpret these facts, if not by admitting :

- That all these sediments (detritus or of plant origin) were deposited in coastal lagoons that were, on rare occasions, invaded by the sea;
- That the bottom of the basin sank as the deposits accumulated, which explains the existence of a considerable thickness of sediment which, however, was all deposited under a shallow layer of water.

Such a tectonic phenomenon, which causes the earth's crust to sink locally, is called *subsidence*.

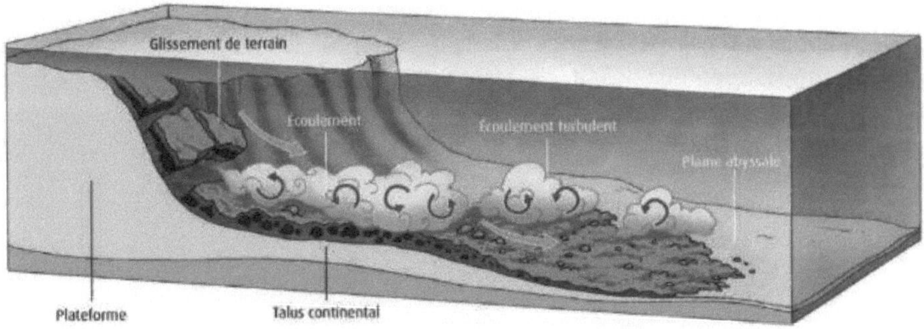

Figure 2.6 *Sedimentation and subsidence*

2.2.2 Horizontal mobility

Until the 1950s, most geologists, apart from a few precursors such as the German astrologer and climatologist Alfred Lothar Wegener, did not accept that the Earth's crust was horizontally mobile; they thought that the continents had been immobile since their formation.

Alfred Lothar Wegener (1880-1930)

They also believed that the oceanic crust remained unchanged because, submerged, it was protected from erosion.

Thus, the ocean floor, which has received the products of continental erosion since the origin of the Earth, must have contained sediments several *billion* years old.

From the 1960s onwards, a veritable revolution in the *Earth Sciences* revealed that the previous facts were *wrong*:

- advances in the study of rock magnetism provide irrefutable evidence of *continental drift*.
- Progress in oceanography proves that the *oceanic crust* is *neither old* (230 million years at the most) *nor inert* (because it is constantly moving).

2.3 OCEAN FORMATION AND CONTINENTAL DRIFT

2.3.1 Wegener's theory

From 1912 onwards, *Alfred Wegener* developed the theory of "*continental drift*". In this theory, the continents "float" on the mantle, which has a higher density, and are driven by translational movements.

At one time, all the continents would have been united into a single

landmass, *the Pangea*, and then separated.
The hypothesis of relative movement between continents was not new and had already been discussed, notably by Darwin.

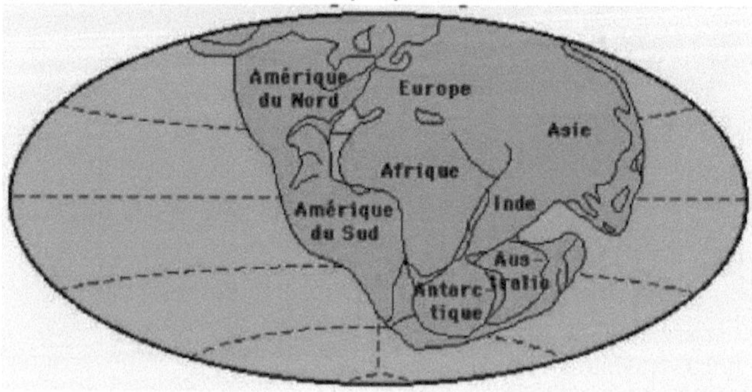

Figure 2.7 *The Wegener Pangea*

2.3.2 Wegener's arguments

Wegener put forward a number of arguments, including the following:

- on the concordance of the coasts on either side of the Atlantic, especially in South America and Africa, whose contours fit together almost perfectly (*morphological evidence)*.
- on the *paleontological* and *geological* similarities between Brazil and South Africa.
- on the *particular flora* that existed during the Permian, i.e. before the *Pangea* broke up, in South Africa, South America, India, Antarctica and Australia ("*Gondwana Continent*"), but not in the rest of Eurasia and North America ("*Angara Continent*").
- on the continuity that exists across the Atlantic between the *ancient mountain ranges* of North America and Europe.
- or the different *climates* recorded by the continents over geological time.

Wegener's theory found a majority of detractors in the geological community.

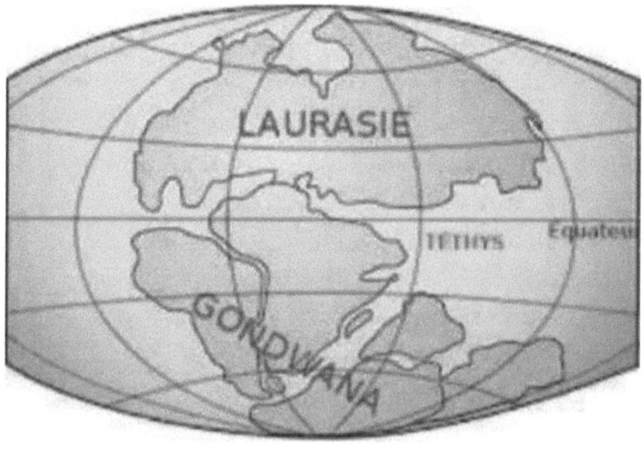

Figure 2.8 *Continental drift, 200 million years ago in the Triassic period. Laurasia in the north and Gondwana in the south.*

2.3.3 East African fault field

East Africa is the most powerful fault field in the world. It is divided into two main parts: to the north, the area around the *Dead Sea* and *Lake Tiberiade*, extended by the *Red Sea* and the *Gulf of Aden*; to the south, the *Rift Valley* region dotted with large lakes.

These *faults* delimit collapsed zones, as shown by the *depth of* the great lakes (the bottom of *Tanganyika* is -700 m below sea level, that of the *Dead* Sea -800 m, the upper level of its waters being -400 m).

Thousands of metres of continental, fluvial and lacustrine sediments have accumulated in the *rift valleys*, along with volcanic products (lava flows and ash).

The presence of very powerful *volcanoes* (Kilimanjaro reaches 6,000 m, Mount Kenya 5,200 m) and the existence of significant seismic activity reflect the intense activity of this continental zone.

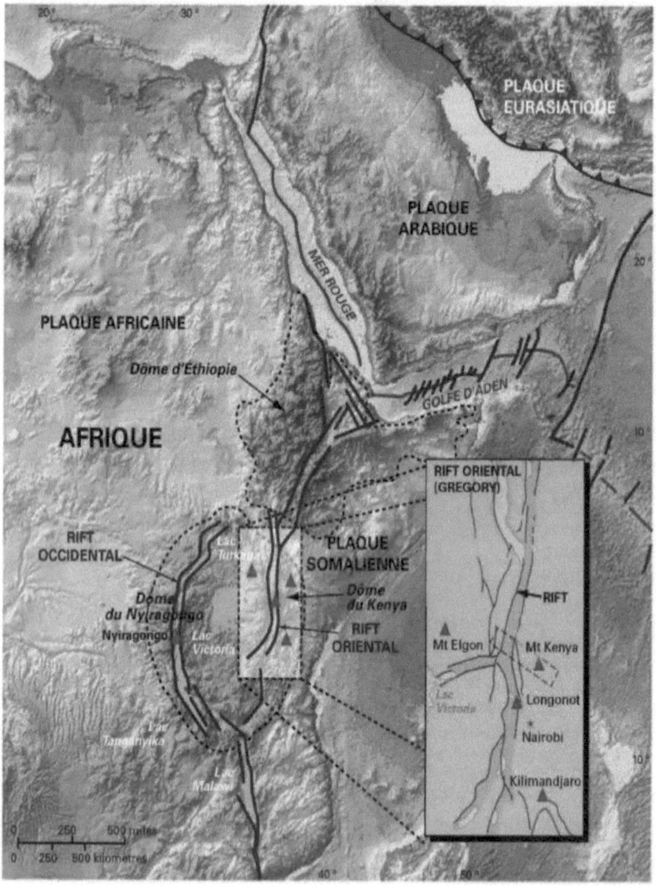

Figure 2.9 *The East African Rift (Source: Geophysics of the Rift-IRD Edition).*

The fracturing of the Earth's crust *that can be observed* at the level of collapse trenches is due to distension forces that tend to pull two continental blocks apart.

A large fissure stretching for several kilometres recently appeared in south-west Kenya.

Figure 2.10 *The fault in Kenya (Source: Le Parisien).*

The crack, which is still growing, has caused part of the Nairobi-Narok motorway to collapse. The crack was originally caused by tectonic activity along the East African Rift, and there is a risk that it will spread as far as Zimbabwe and divide Africa into two continents.

The *East African Rift Valley* stretches for more than 3,000 km from the Gulf of Aden in the north to Zimbabwe in the south, dividing the African plate into two unequal parts: the Somali and Nubian plates.

Activity along the eastern branch of the Rift Valley, which borders Ethiopia, Kenya and Tanzania, became evident when the great fissure suddenly appeared in south-west Kenya.

The division is a very slow process, with an average speed of 2.5 to 5 cm per year, dividing Africa into two continents, without anyone noticing.

And as an explanation, the Earth's lithosphere, formed by the Earth's crust and the upper part of the mantle, is divided into a number of tectonic plates.

These plates are not static, but move relative to each other at varying speeds, 'sliding' on a viscous asthenosphere. It is these tectonic movements that are probably at the origin of this fault.

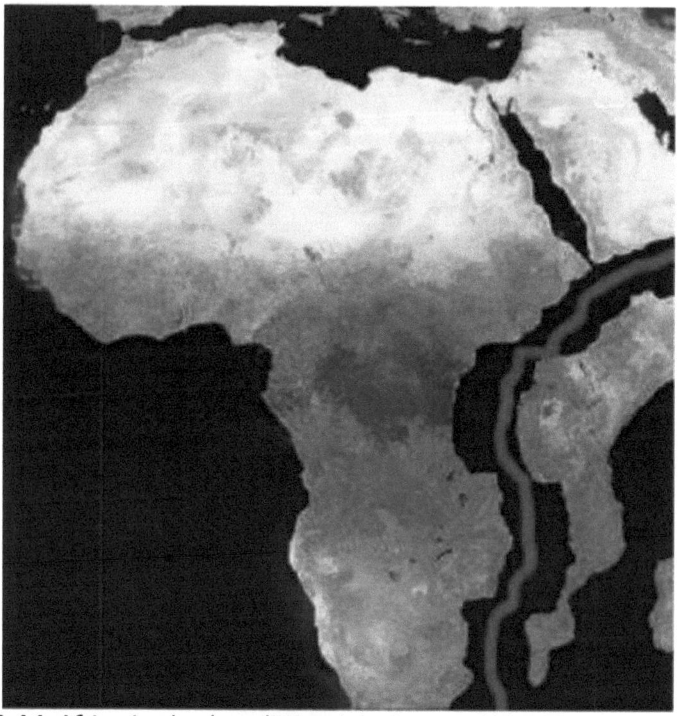

Figure 2.11 *Africa is slowly splitting into two continents*

REVISION QUESTIONS ON CHAPTER 2

1. Explain the origin of earthquakes and seismic waves.
2. Define these terms: epicentre of an earthquake, focus or hypocentre, seismometer, seismograph, paleomagnetism, Gondwana, Wegener Pangea, Laurasia, fault.
3. How important is the Richter scale?
4. Explain the importance of studying seismic waves.
5. What is the difference between isostasy and subsidence?
6. Describe Alfred Wegener's theory of continental drift.

CHAPTER 3

THE VOLCANS

3.1 VOLCANISM, A NATURAL DISASTER

Volcanism is the cause of some of the most spectacular natural disasters in the world. Recent volcanic eruptions have caused numerous deaths, destroyed towns, transformed landscapes, altered the climate and left a lasting mark on people's minds.

Volcanic hazards are significant. Violent explosions, lava flows, glowing clouds and plumes of ash are the much-feared attributes of volcanoes. The formation of volcanoes and their evolution is fascinating.

Usually located at the meeting point of tectonic plates, volcanoes sometimes erupt, causing earthquakes, collapses or tsunamis. However, the landscapes of volcanoes never cease to intrigue us, so unusual and exceptional are they.

3.2 THE WORLD'S MOST DANGEROUS VOLCANOES

Here are the 15 fearsome volcanoes that have been in the news over the past few centuries, decades, years or recently.

3.2.1 Mount Vesuvius (Italy)

Vesuvius is one of the best-known volcanoes in the world... and one of the most dangerous. Although it was responsible for burying the town of Pompeii in 79 BC, eight major explosive eruptions have been recorded over the last 17,000 years. The most recent eruption was in 1944, with large pyroclastic flows on the eastern, northern and western slopes of the mountain. Around ten thousand people from the surrounding villages were evacuated. Today, at least six million people are thought to live in the surrounding area.

3.2.2 Mount Rainier (United States)

Mount Rainier is one of the turbulent volcanoes in the Cascade Range on the west coast of America. And it is particularly closely watched. If a major eruption were to occur, the ice and snow on the summits would mix with the volcanic rocks, forming a gigantic mudflow (lahar). This rapid flow could bury several cities, including neighbouring Tacoma and parts of Seattle. More than two million people could be affected.

3.2.3 Novarupta (United States)

In 1912, the Katmai National Park (Alaska) was the scene of the largest eruption in the world in the 20th century. Almost thirty cubic kilometres of ash and debris were expelled into the air, damaging the Valley of Ten Thousand Smokes... and leaving behind the Novarupta volcano - "new eruption", in Latin. At the time, the region was little exploited. However, if such an event were to happen again, many industries (aviation, fishing, tourism) and wildlife would suffer.

3.2.4 Pinatubo (Philippines)

After Mount Novarupta, Pinatubo is responsible for the second largest eruption of the 20th century. In 1991, the stratovolcano in the west of the Philippine island of Lu^on produced impressive pyroclastic flows, despite having no known history of explosive eruptions. At least 722 people were killed, and a caldera now fills the upper part of the volcano's cone. Today, it is estimated that 21 million people live within 100 kilometres of the potential danger.

3.2.5 Mount Saint Helens (United States)

The eruption of Mount St. Helens on 18 May 1980 in the Cascade Range was the deadliest and most destructive in the country. After a long rest and two months of activity, responsible for the occurrence of 10,000 earthquakes of varying intensity, a huge pyroclastic flow decimated the fauna and flora on its slopes within a radius of 10 kilometres, and sometimes up to 30 kilometres away. Fifty-seven people died. Although the volcano is now considered dormant, its activity is closely monitored,

3.2.6 Agung or Mount Agung (Indonesia)

Explosive eruptions from Agung in central Indonesia are rare but quite powerful. In 1963, intense volcanic activity lasted for more than eleven months, producing ashfalls and pyroclastic flows that caused enormous material and human damage.

The one in 2017 led to the evacuation of 75,000 people within an eleven-kilometre radius - the lower slopes of Agung are densely populated due to the fertility of the volcanic soil. The ash plume was observed throughout 2018.

3.2.7 Mount Fuji (Japan)

The last time the famous Mount Fuji erupted was in 1707, sending debris all the way to Tokyo. Since then, however, the 2011 earthquake on the Pacific coast of Tohoku, responsible for a magnitude 9.1 quake, is believed to have caused a rise in pressure at the highest point of Fujisan.

Experts are therefore warning that the volcano could erupt again. Although it is currently classified as active with a low eruption risk, such an event could affect more than 25 million people in the surrounding area.

3.2.8 Merapi (Indonesia)

With an average eruption every five to ten years, the Merapi volcano on the island of Java is considered to be one of the most active... and dangerous in Indonesia.

This "mountain of fire" also rises above one of the most densely populated regions in the world, with 1,400 inhabitants/km2 on its western and southern slopes. In 2010, during three months of terror, its numerous fiery clouds caused more than 400,000 people to flee their homes, killing 367. Its most recent eruption, of the phreatic type, took place in 2018.

3.2.9 Kilauea (Hawa'i)

Kilauea is known as the most active volcano in the world. A so-called "shield volcano", it produces effusive eruptions that give rise to flows of

fluid lava, occasionally overflowing along the caldera - causing fire hazards and sulphur dioxide emissions.
But it is not impossible for it to become explosive through a *phreatomagmatic* phenomenon: the magma descends below the water table and comes into contact with water, and the rocks are propelled outwards by the steam.

3.2.10 Popocatepetl (Mexico)

Popocatepetl, the "smoking mountain" in Nahuatl, has been active since the end of the 20th century. Nicknamed "Popo" by the locals, it has been experiencing intense outbursts of fumarole, ash and volcanic material for many years, sometimes accompanied by low levels of exposure inside the crater and seismic tremors. Its peaks of activity are closely monitored, however, because a more violent eruption could threaten several tens of millions of people nearby, if it spreads as far as Mexico City or Puebla.

3.2.11 The Yellowstone caldera (United States)

Almost 630,000 years ago, a violent eruption shook what is now Wyoming. Some 1,000 cubic kilometres of rock and ash were ejected, creating a 65-kilometre-wide volcanic depression known as the Yellowstone caldera.
Now dormant, the super volcano hides an imposing underground magma chamber that fills up over the years, making it potentially explosive in the blink of an eye. The park's hydrothermal springs are the result of the presence of this magma at shallow depths.

3.2.12 Nyiragongo (Democratic Republic of Congo)

It is considered to be the most active volcano in Africa, hosting the world's largest near-permanent lava lake. The Nyiragongo stratovolcano is located near Goma, in the Virunga range in the densely populated eastern part of the Republic of Congo (DRC).
If it erupts, it poses a threat to almost two million inhabitants. The last eruption, on 17 January 2002, dumped between 15 and 25 million cubic metres of lava on Goma, killing hundreds of people and devastating the town and its houses.

3.2.13 Cotopaxi (Ecuador)

Like most of the volcanoes in South America's Ring of Fire, Cotopaxi is a particularly dangerous explosive volcano. Its summit, at 5,897 metres above sea level (comparable in height to Kilimanjaro), is covered in snow and ice, giving rise to fears of lahars in the event of eruption.
In 2015, it awoke from its slumber to spew a column of ash eight kilometres high. As a result, residents of the nearby capital Quito, 50 kilometres away, were fitted with masks to protect them from the ash fallout.

3.2.14 Krakatoa (Indonesia)

In August 1883, the volcano Krakatoa (or Krakatau) produced one of the most devastating eruptions in history. Rains of burning ash, waves over fifteen metres high... Some 36,000 people living along the Indonesian Sunda Strait were killed. Forty-five years after this cataclysm, a new mountain rose from the sea in the same place, Anak Krakatau ("Krakatau's child"). A site that is still active, since it was the source of the Sunda Strait tsunami in 2018, as well as an ash cloud in 2022.

3.2.15 Or^fajokull (Iceland)

Or^fajbkull - from the Icelandic "br^fi", "sorry land" - is the country's highest peak and one of Europe's most powerful volcanoes. It is a caldera, occupied by a glacier of the same name. Although it did not appear to have given any sign of life since 1727, geologists have detected increasing seismic and geothermal activity in the area since 2016.

The cause? Magma movements, still relatively weak, but similar to the 1998 activity of the Eyjafjallajbkull volcano - which erupted twelve years later, in 2010.

3.3 FORMATION OF VOLCANOES

Volcanoes form when ash, dust and magma (a mixture of molten rock and gas from inside the earth) rise and accumulate above the earth's surface, known as the "crust". The magma expelled from the volcano is called lava.

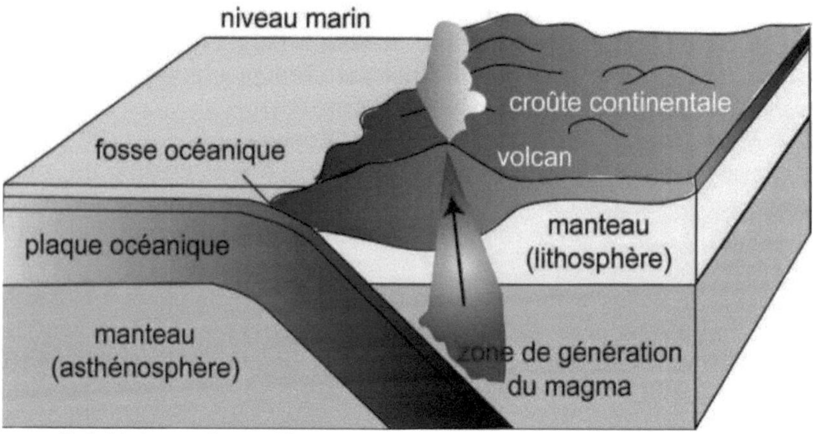

Figure 3.1 *Volcano formation*

3.3.1 Characteristics of eruptions

Active volcanoes in subduction zones are among the highest peaks on Earth. Those in the Andes often exceed 6,000 m in altitude.

A. Eruptive forces

The eruptive styles of these Andean volcanoes vary, but they are all explosive. During explosions, the magma breaks up into ash, lapilli (the size of a pea), slag (the size of an egg), bombs and pumice flows. The explosive momentum is often accompanied by flaming showers and the growth of a viscous lava dome at the bottom of the crater.

B. Explosive eruptions

The explosive nature of eruptions is linked to several factors:

- The chemical composition of the magma, which determines its viscosity
- The amount of gas dissolved in this magma
- The quantity of water that impregnates the volcanic cone (in the case of phreatic explosions)

Magmatic gases escape easily and almost without explosion from very fluid lavas. On the other hand, they can only escape with difficulty from viscous magmas, so they are forced into the latter until the pressure becomes great enough to cause the liquid rock to burst. So the intensity of the explosion depends on the volume of gas and the pressure.

Phreatic explosions are caused by the sudden vaporisation of water on contact with very hot magma.

3.3.2 Characteristics of volcanic rocks

The volcanic rocks of subduction zones display a remarkable range of common mineralogical characteristics. The microlitic structure of the rocks shows that the magma cooled in two stages.

- Very slow cooling in the magma reservoir located in the continental crust at highly variable but generally significant depths (10 to 30 km);
- Rapid cooling on reaching the surface.

3.3.3 Chemical composition

The volcanic rocks emitted in subduction zones are mainly andesites and dacites, which have a chemical composition close to that of the continental crust. Figure 3 shows that andesites are richer in silica than basalts. The viscosity of a magma increases with its silica content.

Magma forms at a depth of 80 to 100 km at the upper limit of the plunging plate. The melting of rocks in the overlying continental lithosphere is due to the water released by the material of the plunging plate. Deeper down (100 to 300km), magma is produced by partial melting of the oceanic crust. In addition, sediments of continental origin, carried along with the plunging plate into the asthenosphere, contribute a small percentage to the generation of Andean magmas.

3.3.2 Fractional crystallisation in the magma reservoir.

The chemical composition of the magmatic 'liquid' changes in the magma reservoir as a result of fractional crystallisation, similar to that observed at rift level. The first crystallisation of certain minerals modifies the composition of the residual liquid. In this way, during successive eruptions, different rocks can be formed from the same primary magma.

In contrast to the plate boundary volcanism we encountered earlier, intra-plate volcanism can occur either in an oceanic plate (e.g. the island of Reunion) or in a continental plate (e.g. the Chaine des Puys in the Auvergne).

3.3.3 Andesitic and basaltic volcanism

There are two main types of volcanism, andesitic volcanism and basaltic volcanism, within the framework of plate tectonics.

A. Andesitic volcanism

Andesitic volcanism is essentially located in subduction zones (convergent lithospheric boundaries) and is characterised by the emission of lavas whose chemical composition corresponds to the calc-alkaline range.

B. Basaltic volcanism

Basaltic volcanism has two distinctly different locations:

- At the oceanic ridges (divergent lithospheric boundaries) the lavas emitted are mainly tholeiitic basalts whose chemical composition falls into the sub-alkaline range (low alkali/silica ratio);
- At hot spots (intra-plate, oceanic or continental volcanism), the basalts emitted have a chemical composition that, unlike the previous ones, is characterised by a much higher alkali/silica ratio, hence the name alkaline basalts.

Table 3.1 *The chemical composition of three lavas.*

	Alkaline basalt	Andesite	Tholeiitic basalt
SiO_2	47,7	57,5	50
^2^2	15,2	16,6	15
$Fe\ O_{22}$	2,3	2,6	3,8
FeO	8,7	4	7,4
MgO	9,7	4,2	7
Cao	8,9	8,7	10,2
$Na\ O_2$	2,7	4,4	2
$K\ O_2$	1,5	2,0	0,3

C. The origin of magmas

It is now accepted that magmas are formed by the partial melting of a pre-existing solid material.

The magmas that *feed* alkaline *volcanism* (intra-plate) form at great depth, in sectors of the Earth's mantle where the temperature is normally high; these are the hot spots. It is estimated that the partial melting rate of mantle peridotites that give rise to alkaline magmas is around 5%. This low melting rate produces a liquid whose composition is markedly different from that of the solid from which it was formed.

In tholeiitic volcanism (oceanic ridges), the magma also comes from the mantle, but in this case the melting rate is high (15 to 20%), which explains the difference in chemical composition between the lavas of the alkaline series and those of the tholeiitic series.

The case of *Andesitic volcanism* (calc-alkaline) posed a problem because, as it occurred in an area where an oceanic plate and a continental plate collided, it was difficult to determine whether the magma came from the melting of one plate, the melting of the other, or the simultaneous melting

of both.
Today, it is considered that :
- The melting of the *plunging plate* is at the origin of the Andesitic magmas.
- The melting of the *continental plate* is more likely to be the source of granitic magmas.

D. The rise of magmas.

The magmas that form volcanic rocks are all *very hot* magmas (1100 to 1300°C), which gives them a lower density than the solids from which they are derived. This low density, combined with low viscosity, gives them high vertical mobility. So, whether magmas are produced in higher zones (ridges) or at greater depths, they quickly reach the surface of the globe without cooling significantly (the temperature of basaltic lava at the time of emission is between 1100 and 1200°C.

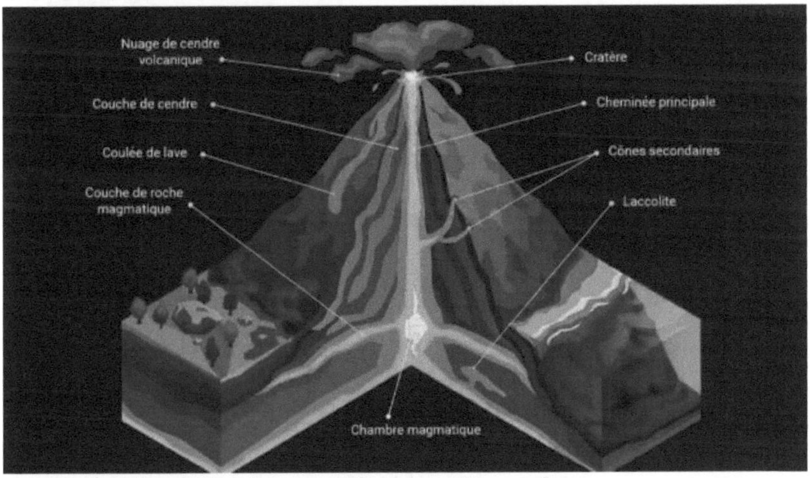

Figure 3.2 *Schematic diagram of an erupting volcano.*

3.4 THE VOLCANOES OF THE DRC

The six dormant or extinct volcanoes are : Muhabura, Gahinga, Sabinyo, Visoke, Karisimbi and Mikeno.

Which *volcanoes are active* in the DRC?

The *Nyiragongo* and *Nyamulagira* volcanoes are the two active volcanoes in the region, known for their Hawaiian-style eruptive dynamics and some of the widest lava flows on earth.

The six extinct volcanoes are : Muhabura, Gahinga, Sabinyo, Visoke, Karisimbi and Mikeno.

3.4.1 THE NYAMLAGIRA VOLCANO

Nyamuragira, also known as Girungo-Namlagira, Namlagira, Nyamlagira or more frequently Nyamulagira, is a volcano in the Democratic Republic of Congo.

Figure 3.3 *Nyamlagira volcano, distant view.*

The most active volcano in Africa, it is capable of producing lava flows over thirty kilometres long that reach Lake Kivu.

Figure 3.4 *The erupting Nyamlagira volcano.*

3.4.2 THE NYIRAGONGO VOLCANO

Nyiragongo, formerly known as *Niragongo*, is a stratovolcano1 rising to an altitude of 3,470 m in the Great Rift Valley in the east of the Democratic Republic of Congo.

It is located in the Virunga mountains, some twenty kilometres north of the town of Goma and Lake Kivu, and west of the Rwandan border.

Nyiragongo is one of the most active1 and dangerous volcanoes in Africa, due to its proximity to densely populated areas, its frequent eruptions, the most recent of which began on 17 May 20023 - with a new eruptive phase on 22 May 2021 - and the presence of a virtually permanent lava lake (a rarity in the world) that can pour down its slopes in long lava flows considered to be the fastest in the world4.

Nyiragongo is located in East Africa, in the extreme east of the Democratic Republic of Congo, less than 10 km from the Rwandan border in the south-east6 and 10 km from the Ugandan border in the north-east.

Part of the Virunga Mountains, a volcanic chain straddling the three States, it is one of the volcanoes of the western branch of the Great Rift Valley and is included in the Virunga National Park7,6.

It is surrounded by Lake Kivu, the Congolese town of Goma and the Rwandan town of Gisenyi to the south, the Nyamuragira volcano to the north-north-west and the Karisimbi and Mikeno volcanoes to the east-north-east6.

The mountain rises to an altitude of 3,470m and has a summit in the shape of a crater 1.2km in diameter1. Its steep slopes, typical of a stratovolcano, are interrupted by the presence of two ancient volcanoes, Baruta to the north and Shaheru to the south, as well as a chain of eruptive vents. These have formed around a hundred volcanic cones stretching along a north-east-south-west axis from the east of the summit to Lake Kivu1.

Figure 3.5 *The erupting Nyiragongo, near Goma in the DRC.*

3.5 THE DRC'S GEOLOGICAL RESOURCES

In terms of its soil and subsoil, the DRC is a "geological scandal", with its vast and diverse mineral resources: copper, cobalt, gold, zinc, diamonds, iron, uranium, niobium, tin, tantalum, tungsten and rare earths, as well as

hydrocarbons (oil and methane gas from Lake Kivu).
The DRC also has significant hydroelectric potential, notably the Inga dam.

3.5.1 MINING RESOURCES IN THE DRC

The DRC is the world's leading producer of cobalt, a strategic raw material for the automotive industry. It is also a major producer of copper (Africa's leading producer) and gold.

A. IRON

Iron is one of the group VIIIb transition elements. After aluminium, iron is the most abundant metal in the Earth's crust (6.2%). It is found in the form of: hematite (Fe_2O_3), siderite ($FeCO_3$) and magnetite (Fe_3O_4).
Pure iron is a grey metal, not very hard. It is essential for living beings.
It reacts with HCl: Fe + 2 HClFeCh + H_2
It reacts with H2SO4: 2 Fe + 3 H_2SO_4 $Fe_2(SO_4)_3$ + 3 H_2
With HNO_3, the iron becomes "passive" as a layer of Fe_3O_4 forms on its surface.
In moist air, iron rusts, forming a porous layer of iron(III) hydrate: $2Fe_2O_3 \cdot 3H_2O$ or limonite.
The oxidation stages of iron are +2 and +3.
Fe (II) compounds :
FeO (black)
$FeSO_4 \cdot 7 H_2O$ (green)
$FeCl_2$ (yellow) FeS (black)
$FeCO_3$ (siderite or iron oligite)
Fe (III) compounds :
Fe_2O_3 (reddish brown, oligistic iron or hematite)
$FeCl_3$ (black)
Iron is the most widely used metal, in the form of iron-carbon alloys, cast iron (2 to 5% C) or steel (0.1 to 1.5% C). It is used in the manufacture of railway rails, structural beams and wire mesh. Nails, screws, electromagnets, dynamos, transformers, etc.
In organic chemistry, iron is used as a catalyst.

Figure 3.6 *Iron ore*

B. COPPER

Copper is not very abundant (0.007% of the earth's crust). It is found in its native state and in a combined state (chalcopyrite: $CuFeS_2$, reddish in

colour, chalcosine Cu_2S or in oxide form, the most important of which is malachite: $CuCO_3$. $Cu(OH)_2$, produced in the DRC, Katanga and Mexico).
After silver, copper is the best conductor of electricity and heat. It is used in alloys, electrical cables and coins.
The DRC has large copper deposits, mainly concentrated in the Katanga provinces. The Albert ALOMBA mine project, Introduction a la Geologie.
Kamoa is one of the world's largest copper resources.
Copper reacts with concentrated hot H2SO4:
$Cu + 2H_2SO_4\ CuSO_4 + 2\ H_2O + SO_2$
It reacts with HNO3 :
$Cu + 4\ HNO_3\ Cu(NO_3)_2 + 2\ H_2O + 2\ NO_2$
The oxidation stages of copper are +1 and +2.
Copper compounds (I): colourless except for Cu_2O (red)
Copper (II) compounds: they are paramagnetic and coloured. For example: CuO (black), CuSO4 (blue), CuS (black).
Copper (II) hydroxide dissolves in ammonia to form a blue complex: $Cu(OH)_2 + 4\ NH_3\ [Cu(NH_3)_4]^{2+} + 2\ OH^-$.

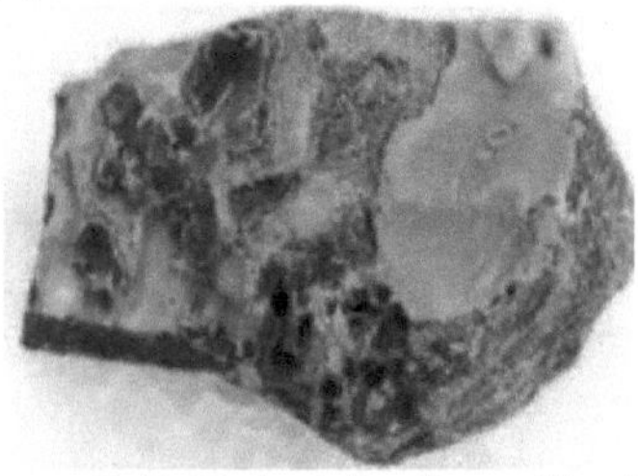

Figure 3.7 *Malachite from the DRC*

C. URANIUM
Uranite is the main ore of uranium and radium and owes its name to uranium. It is a uranium dioxide, but in nature it is more or less an oxide, so its composition varies between two poles: VO2 and UO_3.
Uranium is part of the composition of the lithosphere and in different forms, such as pitchblende, which contains between 40% and 90% U3O8 oxide in the DRC and Canada, carnotite or potassium uranium vanadate (Colorado) and chalcolite or copper uranium phosphate (Portugal).
Uranium is produced in many countries, particularly in North America, Africa and Asia. The main producers are Kazakhstan, Canada, Australia, Niger (the main African producer) and Namibia, which account for almost 80% of world production.
The DRC is one of Africa's uranium-producing countries. The Shinkolobwe mine in Katanga produces uranium and cobalt. Uranium from Katanga was used by the Americans to make the atomic bomb they dropped on Nagasaki and Hiroshima on 6 August 1945.

Figure 3.8 *Uranium ore*

Uranium has a metallic appearance similar to that of iron; it is paramagnetic and has the properties of a transition element. It forms oxygen cations such as UO^{2+} and UO_2^{2+} as well as numerous derivatives (chlorides, sulphates, oxides, etc.).

Preparing uranium involves transforming the ore into an oxide, which is then reduced by carbon at the temperature of the electric furnace.

When freshly prepared, uranium emits only alpha radiation and then a little beta radiation.

D. THE COBALT

Cobalt with the electronic structure [Ar] 4s2 3d7 is the second element of the eighth secondary group, this transition metal is part of the iron group. The simple cobalt body has physical properties quite similar to those of iron and nickel. Chemically, it is less reactive than iron. Cobalt is also a Group 9 element, the first three of which - Co, Rh and Ir - make up the cobalt group.

Cobalt is a metal that is particularly in demand in green technologies driven by the energy transition. 80% of cobalt is consumed in the manufacture of lithium-ion batteries.

Cobalt is co-produced by the extraction of copper ores in the Democratic Republic of Congo and Zambia, the extraction of nickel sulphide ores in Russia, Canada and Australia, and lateritic nickel ores in New Caledonia, Cuba, Indonesia, Australia, the Philippines and Madagascar.

By 2021, the Democratic Republic of Congo was the world's leading cobalt miner, with production of 120,000 tonnes.

Figure 3.9 *Cobalt*

E. GOLD

Along with silver and platinum, gold (symbol: Au) is one of the so-called noble metals, which are unalterable in air. They are very expensive and less abundant in nature.

Gold is a soft, very dense, bright yellow metal, the most ductile and malleable of all metals.

Gold forms two series of compounds: aureous compounds (Au^{+}) and auric compounds (Au).$^{3+}$

In particular, it forms relatively unstable salts: when heated, gold (III) chloride AuCl3 is first transformed into gold (I) chloride AuCl, then into metallic gold.

AuCl3 AuCl + Clz^ and 2AuCl 2Au + Clj".

Gold is used for gilding, either by applying gold leaf or by electrolysis to make jewellery. Gold is used to make alloys, in particular with copper, silver and nickel (the gold content of an alloy is expressed in carats, i.e. the number of parts of gold per 24 parts of alloy; one carat is worth 0.205 grams; pure gold is 24 carats).

White gold is a white alloy of gold and nickel used in jewellery. Gold is also used to make dental prostheses and scientific equipment.

The main gold mines are located in China, Australia, South Africa, Russia, Peru and Canada.

The DRC is a major African gold producer, and its first gold refinery is located in Bukavu, South Kivu. In the first quarter of 2022, total gold production is estimated at 13,107.33 kg.

Figure 3.10 *Gold*

F. LE COLTAN

Coltan (short for columbite-tantalite) is a black or reddish-brown ore from which *niobium* (originally called columbium) and tantalum are extracted. It is formed by combining two minerals, *columbite* [(Fe, Mn) Nb_2O_6] and *tantalite* [(Fe, Mn) (Ta, Nb)$_2O_6$], in variable proportions.

Coltan was discovered in New England (USA). Coltan is found in commercial quantities in Central Africa, in particular in the Democratic Republic of Congo in the Kivu region, which holds between

60 and 80% of the *world's reserves,* as well *as* in Australia, Brazil, Canada, Spain and China.
Coltan is an important source for the production of tantalum and, to a lesser extent, niobium. Tantalum is highly prized for its high resistance to corrosion. Considered a *strategic metal*, it is mainly used in the manufacture of capacitors for electronic equipment, but is also used in the composition of cobalt and nickel alloys in aeronautics, particularly in the manufacture of jet engines. It is also used as a coating in heat exchangers and in alloys for cutting and turning tools.
Tantalum, prepared from coltan, is essential for the manufacture of electronic components, notably capacitors and surface wave filters, used in particular in mobile phones. The electronics sector thus monopolises 60 to 80% of the tantalum market.
In 2000, a worldwide shortage and a rise in tantalum prices prevented Sony's *PlayStation 2* from being manufactured in sufficient quantities.
Coltan mine in Central Africa, Coltan is at the heart of the war in the Democratic Republic of Congo (DRC), one of the deadliest conflicts since the Second World War, with more than 5 million people killed.
An expert report presented to the UN Security Council in 2001 denounced the large quantities of this mineral illegally extracted from the soil of the Democratic Republic of Congo and smuggled, in agreement with Western contractors, by the armies of Uganda, Rwanda, Burundi and the DRC, which occupied Province Orientale and the Kivu region in the east of the country.
The Rwandan army is said to have made 250 million dollars from illegal trafficking in this mineral between 1998 and 2000.
These illegal mining operations encourage foreign troops to stay in Kivu and endanger the region's forest ecosystem.
The three states implicated in the UN report denied any involvement in coltan trafficking.
Because of the difficulty of controlling the legality of mining operations in the region, several electronics manufacturers have decided to "officially" boycott coltan from Central Africa, preferring to source it from Australia.
Rwanda can also produce columbite-tantalite ore, at least in the Gatumba region on the Congo-Nile ridge. Large duckbill crystals were found there by geologists from SOMIRWA (Societe des Mines du Rwanda), which ceased its cassiterite and tungsten activities around 1982.

Figure 3.11 *A piece of colombite-tantalite measuring 6.0 x 2.5 x 2.1 cm.*

G. THE DIAMOND

Carbon comes in two allotropic varieties, one crystalline and the other amorphous. A structural phenomenon by virtue of which a simple body can take several different physical forms.

Amorphous carbon is a form of carbon that has no crystalline structure, or in fact is highly disordered. It is the constituent of a number of substances called coals, containing carbon in varying proportions. Depending on their origin, a distinction is made between natural and artificial coals.

Natural coals come from the decomposition of vegetation accumulated in marecous soils. Depending on the stage of evolution, a distinction is made between peat, lignite, coals and anthracites.

Artificial coals are produced artificially for certain specific uses: coke (fuel and reducer in metallurgy), retort black (manufacture of electrodes), carbon black (manufacture of shoe polish, printing inks), animal black from the calcination of bones (decolourisation of sugar juices, etc.), charcoal, etc.

Crystalline carbon comes in two forms: graphite and diamond.

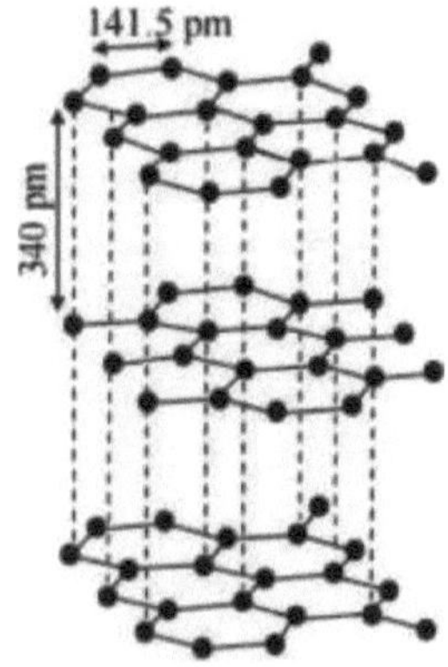

Figure 3.12 *The crystalline structure of graphite*

Natural graphite is one of the crystalline forms of the element carbon. Its atoms are organised in a stack of graphite sheets with a hexagonal structure. Natural graphite is found in metamorphic and magmatic rocks and in meteorites. It has a density of around 2.2 g/cm3 and is greyish in colour. Graphite is used to make pencil leads and carbon paper.
In diamond, each carbon atom is linked to four other atoms, forming a network in all three dimensions of space. This gives the mineral great hardness, making it the hardest mineral known to man. Diamond is highly refractive (refractive index: 2.41), and shines with a bright sparkle - diamond fire.

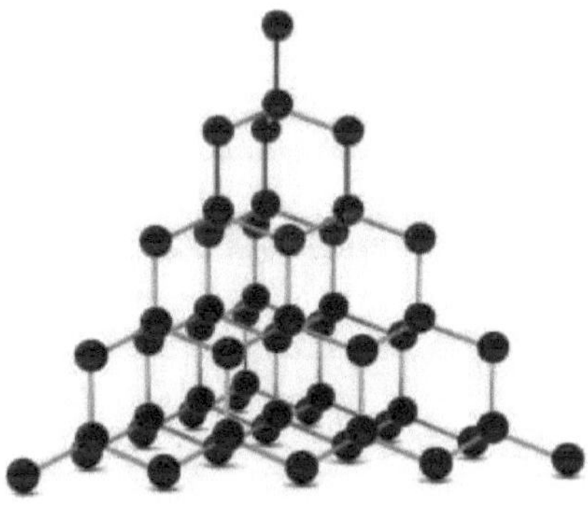

Figure 3.13 *The crystal structure of diamond*

Colourless diamond samples are cut to make precious stones that are highly prized in jewellery. Black diamonds, which are less valuable, are used in the manufacture of cutting tools.
The weight of a diamond, like the weight of precious metals (Pt, Au), is

expressed in carats. A carat is worth 0.205 grams.
Diamond production in the DRC has plummeted over the past 5 years, from 18.9 million carats in 2017 to 12.9 million carats in 2021, according to statistics from the Ministry of Mines. And most Congolese diamonds come from artisanal production, i.e. 71% in 2021.
In the first quarter of 2022, the Democratic Republic of Congo (DRC) produced 4,581,679.13 carats of diamonds, of which 2,376,116.52 carats were mined artisanally (51.86%), 2,201,454.16 carats were mined industrially (48.05%) and 4,108.45 carats were mined semi-industrially (0.09%).
These are Societe Anhui Congo d'Investissement Minier (SACIM) with 2,152,668.73 carats, or 97.78%, and Miniere de Bakwanga (MIBA) with 48,785.43 carats, or 2.22%. North Kivu: 12.58 carats, or 0.001%. According to the CTCPM, total production for the first half of 2022 is estimated at 4,581,679.13 carats.
The price of a 1-carat diamond is generally estimated at between €2,000 and €2,500.

Figure 3.14 *The diamond*

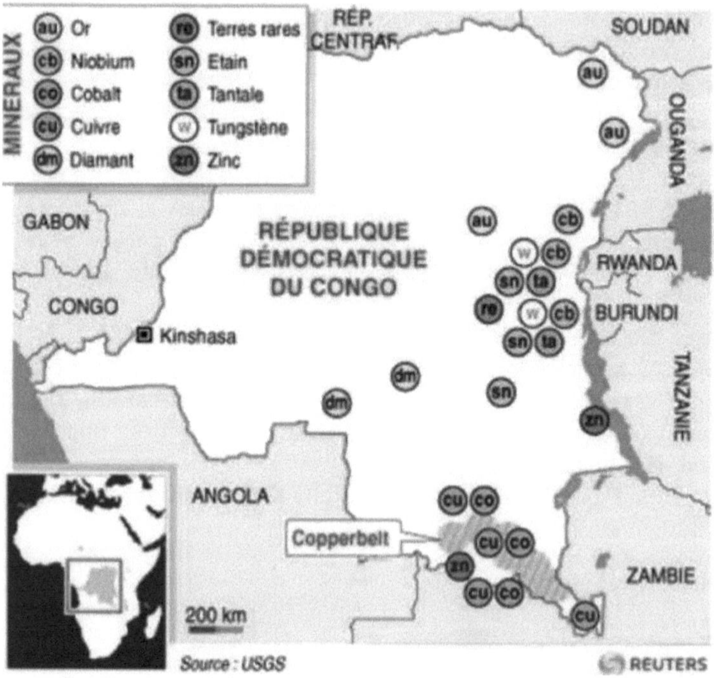

Figure 3.15 *Mineral deposits in the DRC (Source: USGS, Reuters)*

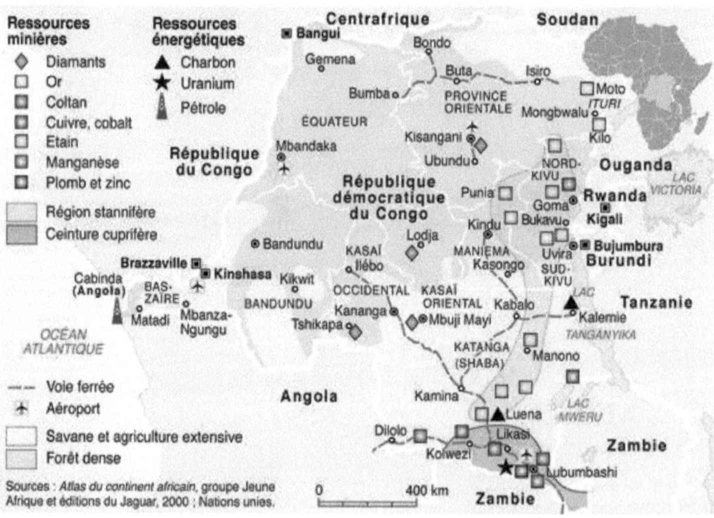

Figure 3.16 *The DRC's mining and energy resources (Atlas of the African continent, Jeune Afrique group and Jaguar edition, 2000, United Nations).*

3.5.2 OILFIELDS IN THE DRC

The *oil blocks* concerned are spread across different regions of the country: three in the coastal basin, nine in the Central Cuvette, eleven in the Tanganyika Graben and four in the Albertine Graben. The three gas blocks are located in Lake Kivu.

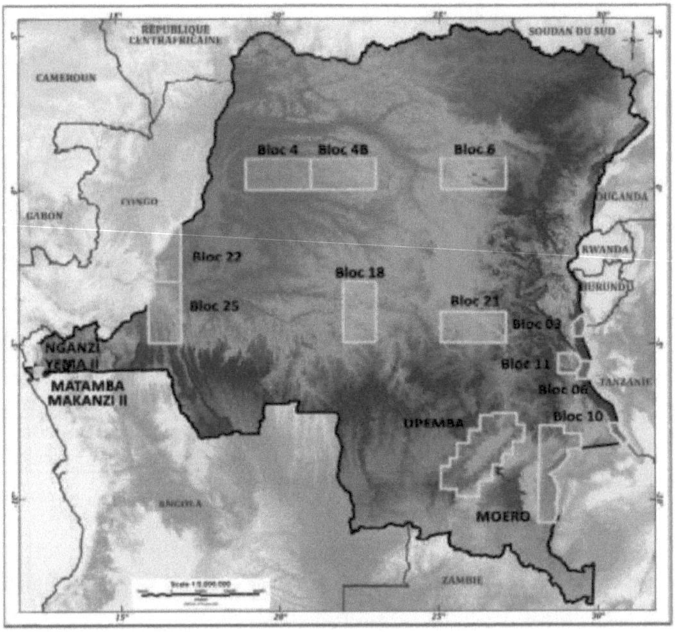

Figure 3.17 *The 27 oil blocks of the DRC.*
(Source: Ministry of Hydrocarbons: Upstream-Petroleum-17)

3.5.3 HYDROELECTRIC POTENTIAL OF THE DRC

The DRC's *hydroelectric potential* is estimated at 100 gigawatts, putting the country in third place in the world behind China and Russia. However, only 2.5% of this capacity has been developed to date.

Eight hydroelectric power stations are in operation, and others are under construction or planned.

Table 3.2 *Lists of hydroelectric power stations in the DRC.*

STATION	CAPACITY (MW)	TYPE	RIVIERE
Inga II power station	1,424	Courses at river	Congo River
Inga I power station	351	Courses at river	Congo River
Power station of Nseke	260		Riviere Lualaba
Ruzizi II power station	45	Reservoir	Ruzizi River
Ruzizi I power station	40	Reservoir	Ruzizi River
Rutshuru hydroelectric power station	13,8	Courses at river	Riviere Rutshuru
Mutwanga hydroelectric	0,4	Courses at	

power station		river	
Mobayi power station		Reservoir	

Table 3.3 *Hydroelectric stations under construction or planned in the DRC.*

Station	Capacity (MW)	Type	Riviere	Status
Power station of Grand Inga	39 000	River course	Congo River	Proposes
Inga III power station	4 800	River course	Congo River	Transfer of markets
Busanga dam	240	Reservoir	Riviere Lualaba	Proposes
Zongo II dam	150	Reservoir		Under construction
Ruzizi III power station	145	Reservoir	Riviere Ruzizi	Under construction
Katende dam	64	Run of River		Under construction
Kakobola dam	10,5		Riviere Lufuku	Under construction

3.6 THE GEOLOGICAL MAP

3.6.1 DEFINITION AND PURPOSE OF A GEOLOGICAL MAP

The purpose of a geological map is to show the nature, age and geometry of the landforms that make up the relief of a region, usually on a previously established topographic base.

This representation is done at different scales, from mining plans, which can be surveyed at 1/1,000^{e} to summary maps published at a very small scale (Geological Atlas of the World, at 1/10,000,000^{e} for example).

3.6.2 IMPORTANCE OF A GEOLOGICAL MAP

Geological mapping serves three purposes: scientific, economic and cultural.

A A scientific necessity

As the geologist's approach is essentially analytical, the geological map is a *graphic translation* of this analysis. Arranging the terrain and the events it has recorded in time and space, it is the basis for any reconstruction of the history of the earth's crust.

On the other hand, the geological map is the indispensable *medium for* many other, more specialised cartographic documents, such as maps showing the relationship between seismic zones and certain faults in the earth's crust (*seismotectonic maps*) or reconstructions of the location of land and sea at different geological periods (*paleogeographic maps*).

B. An economic necessity

A geological map provides the *basic document* for prospecting for deposits of useful substances: energy, ores, construction materials, water. It is also essential for planning the siting of major structures: dams, motorways, nuclear power stations, etc.

C. A cultural need

The importance of the *environment* in everyday life, curiosity about nature and concern for its preservation have led to a growing interest in what surrounds us. Reading geological maps contributes to this cultural development.

3.6.3 DRAWING UP A GEOLOGICAL MAP

Drawing up a geological map is the culmination of work involving many aspects of the earth sciences.

A. First or documentary phase

The first step is to compile existing geological documentation on the area to be studied. This preliminary work includes both a bibliographical study and an analysis of aerial photographs.

B. Second phase or field work

Once the preliminary work has been completed, fieldwork can begin. This will consist firstly of *reconnaissance surveys* to establish the major geological units.

Geological section surveys and initial rock studies (taking samples) will enable us to determine the *age* and *nature of* the terrain encountered.

We then move on to the *detailed survey* stage. In most cases, this detailed work is carried out at a larger scale than that of the definitive map (1:25,000^{e} for the 1:50,000 map^{e} , for example).

Fieldwork requires a constant *change of scale*, from millimetre scale (observations with a magnifying glass) to landscape analysis (several kilometres); observations compatible with the map scale are recorded on the map, while other observations are recorded in the field notebook.

Alternating with the field work, and just as important, laboratory studies will draw on *a wide range of disciplines*. These studies generally take place over several years, enabling us to gain a better understanding of *the nature* (composition and geometry) and *history* of the geological formations that will be represented on the map.

C. Third or synthesis phase

The final stage in the production of the geological map consists of *summarising* the previous studies and transferring them to the *model* map.

The explanations and meanings *of the legends*, *signs* and *traces* enable readers to familiarise themselves with the geological map and make the most of It.

The meaning of geological traces, *signs and symbols* provide additional information, such as *dip signs*.

The map *legend*, and in particular the *coloured insets* on the map sides, the letter *indices* and the numbers, provide information on the *age* and *nature* of the map sets.

The *explanatory note* which accompanies the map describes, in more detail than the legend, the different types of rock and indicates any information which could not be translated graphically (drilling data, old observations from disused mines, etc.). The final map, which will be used by geologists and the general public, will be drawn and printed from the model.

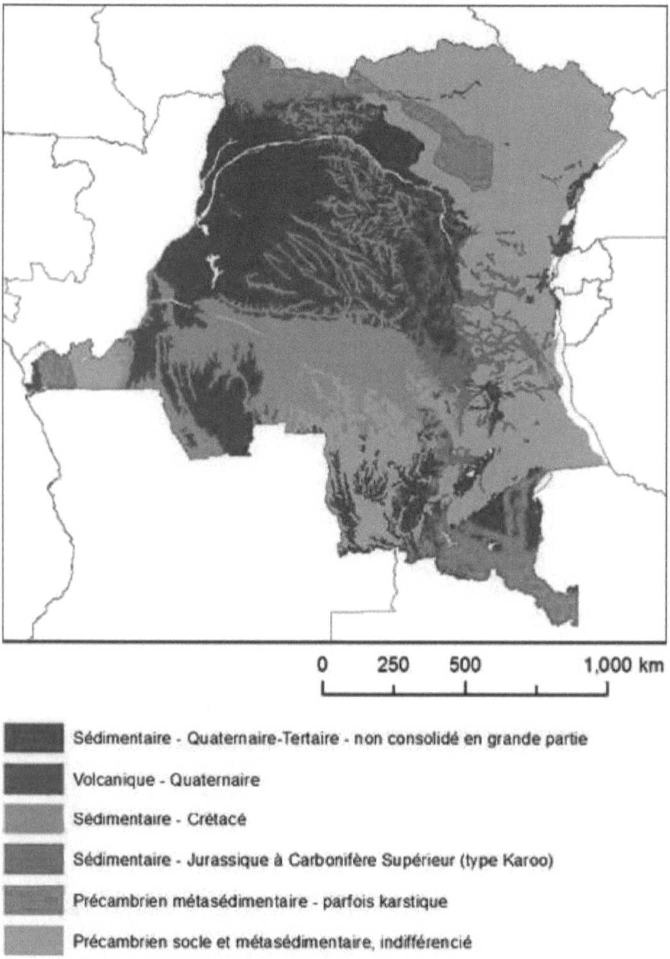

Sddimentaue - Quaternary-Tertiary - not consohd£ for the most part
Volcanic - Quaternary
S£dimentary - Cr£tac6
Sedimentary - Jurassic to Upper Carboniferous (Karoo type)
Prëcaтbпeп metasedimentary - sometimes karst
Prëcaтbпeп base and m^tasedirnentaife. mdiff£renci£

Figure 3.18 *An example of a geological map of the DRC (Source: Furon R and Lombard J, 1964)*

REVISION QUESTIONS ON CHAPTER 3

1. Name 5 dangerous volcanoes in the world that you know of.
2. What are the risks associated with volcanic eruptions?
3. What are the advantages of having a volcano?
4. What is plate tectonics?
5. How do volcanoes form?
6. Explain the characteristics of volcanic eruptions.
7. What are the characteristics of volcanic rock?
8. Explain these concepts: magma, andesitic volcanism, basaltic volcanism, tholeiitic volcanism, a carat, legend of a geological map.
9. Give the names of six extinct or dormant volcanoes in the DRC and two active volcanoes in the DRC. What is the highest volcanic mountain in our solar system?
10. What are the DRC's main geological resources? And what can be done to ensure that these resources benefit the Congolese people?
11. Define a geological map, give its importance and explain how it can be produced.

CHAPTER 4

THE FLOORS

4.1 INTRODUCTION

Soil is a mixture of mineral substances resulting from the decomposition of *bedrock* by *physical and chemical agents*, and specific organic matter produced by the decomposition of plant residues by *biological agents*.

The difference between these two types of material is in fact essentially due to the *'joining' forces* between these grains: in the case of *rocks*, these forces are strong and permanent on a human timescale; in the case of *soils*, they are much weaker and can vary over time.

The *soil* is a highly complex environment populated by an abundant and diverse *fauna* (Coleoptera, Myriapoda, Ants, Collembola, Nematoda, Mites, Rotifera, Protozoa, various larvae and many other microscopic organisms).

This fauna is complemented by an *army* of bacteria, actinomycetes and fungi.

There are also many *abiotic factors* involved in soil ecology, such as water in the soil, soil salinity, soil pH and soil aeration.

The *study of soils* (formation, texture, nature and properties of clays, structure, profiles) is called *pedology*. Roots and organisms are generally forgotten, which is why the study of soils as a particular ecosystem is called *edaphology*.

4.2 SOIL FORMATION

Soil is the surface layer of the Earth. It is formed on Earth by the action of *climate* (temperature and precipitation), *flora* (which provide most of the organic matter and define its evolution), *natural disasters* and *bedrock* (physical properties and chemical composition).

4.2.1 STAGES IN SOIL FORMATION

Soil is the result of the *union of* mineral matter (clays from bedrock) and organic matter (humus from organic debris).

The mineral elements in the soil, welded together by the clay-humus complex, form aggregates between which *water* and *air* circulate.

Broadly speaking, there are *3 stages in* the formation of soil:

A. The first stage is the alteration of the bedrock.

This is the result of numerous *physical* processes (such as frost or root penetration) that fragment the rock, but also of *chemical* processes (attack on the soil by acid-saturated water) that dissolve the limestone and hydrolyse the silicate minerals to create alternating complexes (clay, iron oxides, salts) that cement the grains that resulted from the previous fragmentation.

B. The second stage is the incorporation of the organic matter.

Plant debris (leaves and dead wood in forests, straw or stubble in cultivated land, etc.) is a *source* of organic matter for the soil.

This debris is fragmented in several successive waves. They are then mineralised by releasing chemical elements contained in the various organic compounds (Carbon, Nitrogen, Oxygen...) of the litter under different mineral substances. This process is carried out by bacteria and fungi. The debris is finally humified

C. In the third stage, the horizons are finally distinguished by the action of infiltrating water.

The action of this infiltrated water, known as *leaching*, allows soluble elements to be carried away and accumulated to form *horizons*.

4.2.2 ACTION OF CLIMATE AND VEGETATION COVER

Soil is slowly formed from rock and populated by living organisms; it *develops* and *matures slowly* under the action of the climate and the vegetation that covers it.

Soil *is an ecosystem that evolves slowly* until a state of equilibrium is reached. Soil dynamics are marked by the migration of clay colloids poetised with soluble biogenic elements that accompany the movement of water.

In *humid climates, downward migration* leads to *leaching* and the formation of the A2 leaching horizon, a more clayey horizon (B); sometimes the phenomenon is complicated by *podzolisation* with the formation of the B horizon, which accumulates aluminium and iron sesquioxide and humic matter.

A *soil horizon* is defined as a layer of mineral or organic soil material, approximately parallel to the earth's surface, the characteristics of which are modified by soil formation processes.

In *arid climates, upward migration* leads to the efflorescence of a crust or caliche of solid mineral salts on the surface.

The zonation of the soil's mineral matter into superimposed horizons is complicated by a stratification of the organic matter; depending on the case, thick *litter* and humus (coprogenous or myogenic) are added in more or less well-defined horizons (A0, A1).

Well-drained soils established on horizontal or sloping topography (*plakor*) reflect in their profile the temperature, rainfall and vegetation conditions of the bioclimatic zones where they are found.

We can thus speak of: *zonal soils*, which are contrasted with *interzonal soils* that are conditioned more by particular factors of drainage or mineral wealth than by climate, and non-walled, skeletal *azonal soils* on slightly decomposed bedrock.

4.2.3 SOIL HORIZONS

The soil cover (all the soils that cover the earth) is constantly evolving, developing both from deep bedrock and from surface organic matter.

These transformations give rise to superimposed volumes of mineral and organic matter, more or less parallel to the surface of the ground, known

as *horizons*. Their number varies according to the state of evolution of the soil cover.
Horizons differ from one another in *colour*, *texture*, abundance of *coarse* elements, *spatial* arrangement of constituents (solids, liquids, gases) and *associated voids*. These characteristics, which can be observed with the naked eye, are used by the pedologist during field studies to identify the boundaries of each horizon.
Vertically, the thickness of a horizon varies from a few centimetres to several metres, and its upper and lower limits are more or less clear. *Laterally*, a horizon is transformed into another horizon, over an indefinite distance ranging from a metre to a kilometre.
The superimposition of horizons forms a *soil profile, with* all the horizons in space forming the *soil cover*.
Horizons" are the basis for describing the soil cover. By examining the vertical wall of a pedological pit dug in the ground, the pedologist can divide the *soil cover* into horizons and determine their number.

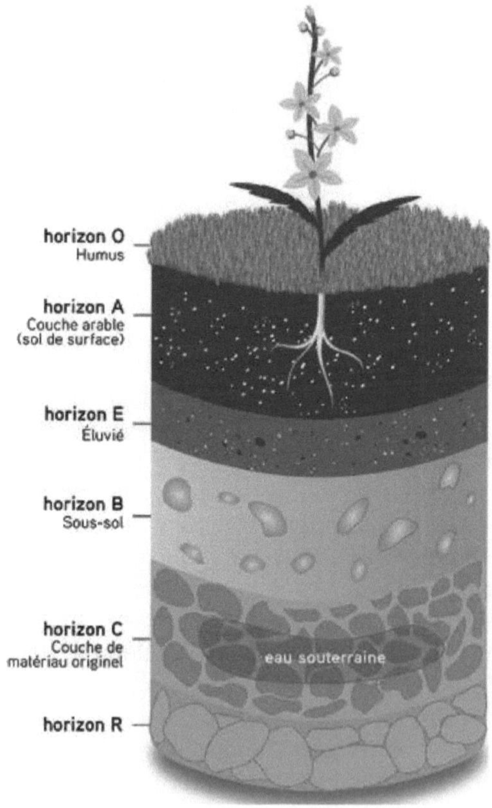

Figure 4.1 *Soil horizons*

4.2.4 DEVELOPMENT AND EVOLUTION OF HORIZONS

The horizons are the result of: the degradation by *fragmentation* and *alteration* of the underlying bedrock under the combined effects of water, air and climate; the transformation and mixing of the products of this alteration (sands, silts, clays) with the fresh organic matter (plant and animal debris) deposited on the surface and then decomposed into humus; the migration, vertical displacement and fixation of certain constituent elements (minerals, clay, iron, organic matter, etc.) over time..) over time.

The order and speed at which the different types of horizons are superimposed vary according to *climatic conditions* (wet, hot, tropical, etc.), *topography* (slopes, plain) and source *rock* (limestone, granite, shale, etc.).

4.2.5 MECHANISMS OF HORIZON FORMATION

These mechanisms work together. Depending on variations in humidity, temperature and biological activity, the physical, chemical and mechanical properties of the constituents are constantly transformed, and mineral, liquid, gaseous and organic matter is transferred.

A. Mechanism of alteration of rocks and their constituents.

In contact with the atmosphere, rainwater and plant and animal life, bedrock and the minerals that make it up degrade and alter to produce soil.

During the alteration phase, some of the minerals (calcite, clay, quartz, etc.) that make up the original rock separate from each other, become poorer, lose or gain chemical elements and are transformed. Other minerals dissolve or are carried away by rainwater, or recombine and crystallise to form new minerals that enrich the soil being formed.

The new minerals migrate through the various layers of soil that are forming and mix with the decomposing altered rock. These minerals are mainly: clays, calcite, iron oxyhydroxides, aluminium oxyhydroxides, etc.

B. Biological mechanisms and the accumulation of organic matter.

The living animal and plant species that develop in and on the altered rock will promote and accelerate the alteration mechanisms: desaggregation, hydrolysis (dissolution), leaching (entrainment), generation of new minerals and contribute to the revolution and functioning of the soil. Dead animal and plant species are transformed into decomposed organic matter (humus).

Organic matter, whose composition varies according to the fauna, vegetation, climatic conditions, bedrock, etc., accumulates and migrates more or less deeply from the soil surface to the bedrock.

C. Mechanism for the release, migration and accumulation of constituents resulting from the alteration and evolution of organic matter.

The constituents produced by the alteration of rocks and biological activities are subject to migratory phenomena: movement of solid particles by gravity, by water entrainment, by movement in soluble form.
These movements take place from top to bottom under the effect of gravity, vertically from bottom to top by capillarity and laterally, from top to bottom of the slopes.
D. Mechanisms of arrangement and aggregation of constituents resulting from alteration, then set up by migration.
These are the complex mechanisms that lead to the creation of colours, aggregates and voids.
4.2.6 NAMING THE HORIZONS
These mechanisms (transformations, migrations, displacements of certain soil constituents) lead to the chronological formation and development of one or more horizons that differ from one another in the presence and proportion of organic matter, mineral matter and altered rock.
The presence in a horizon of certain specific morphological characteristics (constituents, pedological features, structure, colour, etc.) combined with various other elements (position in relation to other horizons, etc.) has enabled pedologists to define more than 70 types of horizon known as reference horizons.
And the reference horizons are denominated by letters: O, A, B, E, S, C, R, M.... according to an international nomenclature.
Here is a description of the main reference horizons.
A. *O horizons (organic horizons)*
They are in contact with the atmosphere and the lithosphere. Organo-mineral, dark in colour, they are essentially composed of more or less transformed, but still recognisable, plant debris.
Thickness varies with the intensity of biological activity and climate. Depending on its thickness, the O horizon is divided into three superimposed layers: OL, or litter composed of undecomposed plant debris; Of, the fermentation layer, where the plant debris is partially decomposed by biological activity; Oh, the humification layer, where the plants have disappeared and are transformed into humus.
Depending on the number of layers present, the words *Mull*, *Moder* or *Mor* are used.
B. *Horizons A (organo-mineral horizons)*
Dark in colour and located below the O horizon, it is made up of plant and animal debris decomposed into humus (humification). It contains both organic and mineral matter.
C. *E horizons (mineral horizons, known as eluvial horizons)*
They are below the A horizon, essentially mineral, leached, depleted in clay minerals, salts, carbonates, hydroxides, light in colour.
D. *B horizons (illuvial mineral horizons)*

Located beneath the E horizons, where elements (clays, iron, aluminium) accumulate, they are enriched in clay minerals, salts, carbonates and hydroxides.

E. *Horizons* Shorizons

minerals

F. *Horizons C*

They correspond to the more or less altered bedrock.

G. *Horizons R, M and D*

R: hard rock; M: soft, loose rock; D: reworked rock that may or may not have given rise to the overlying material.

Each horizon (O, A, E, B, ...) is subdivided into subgroups according to the evolution of pedogenesis and the proportion of certain soil constituents. For example, depending on the importance of the minerals present, the main types of B horizon are as follows: Bt (clay), Bo (iron hydroxides, al), Bh (organic matter), Bca (limestone) ...

4.2.7 SOIL HORIZONS AND DIFFERENTIATION

Depending on the succession and nature of the superimposed horizons, we have soils that are more or less different, more or less evolved. For example, in slightly different soils the succession of horizons is of the type : A-C-R; in moderately different soils, the succession of horizons may be : A- S-C-R and in very different soils the succession of horizons may be : A-E-Bt-S-C-R.

Limestone shale = rock containing silicate minerals and limestone. Alteration of this rock produces sands, silts, clays and limestone.

Stage 1: soil with few differences.

Alteration of the calcareous schist and superficial accumulation of organic matter; the soil is divided into two horizons: an organic horizon and an alteration horizon. The transitions between these two horizons and with the parent rock are gradual; the soil is not very differentiated.

Stage 2: moderately differentiated soil

Alteration continues, the soil deepens, the organic matter continues to accumulate on the surface, the limestone leaves the surface horizons and accumulates at depth. The soil is divided into three horizons: an organic horizon leached of limestone, a horizon leached of limestone, and a horizon of limestone accumulation and shale alteration. The transitions between these three horizons and the parent rock are gradual; the soil is moderately differentiated.

Stage 3: very different flooring

The alteration continued, the soil continued to deepen; the leaching and accumulation of limestone increased, the leaching and accumulation of clay and iron increased, the organic matter became less abundant on the surface as a result of the depletion of clay, the soil became Albert ALOMBA, Introduction a la Geologie.

differentiated into four horizons. These are: an organic horizon leached of limestone, clay and iron; a horizon leached of limestone, clay and iron; a well-drained horizon leached of limestone but accumulating clay and iron; a horizon of limestone accumulation and shale alteration. All the transitions between horizons are gradual. The soil has become very differentiated.

4.2.8 ELEMENTS THAT CHARACTERISE HORIZONS

The colour, texture and abundance of coarse elements and elemental organisations are used to describe and characterise the horizons. We move vertically from one horizon to another by modifying the characteristics of these elements.

A. The colour of a horizon

Colour indicates the physiognomy of a horizon, and can be homogeneous or heterogeneous, varying vertically and laterally.

Soil colours are interpreted in terms of the following constituents and mechanisms:

- *Biological, animal and vegetable activities.*

They accumulate organic matter. Colours darken. The organic matter takes on a darker colour (black, brown, dark grey).

- *The migration and accumulation of minerals (clay, iron, limestone, salts, etc.).*

Horizons depleted of clay and iron become lighter, while those enriched become yellow, brown or red. Horizons rich in limestone or salts turn white. Limestone and soluble salts (chlorides and carbonates) turn white.

- *Water regimes*

Well-draining horizons, which receive a lot of water and dry out quickly, are red; moderately well-draining horizons are brown or yellow; poorly draining horizons are grey or mottled.

Iron colours differently depending on its state: ferrous iron colours grey and blue, due to excess water; goelite colours brown and yellow FeO(OH) in the presence of constant humidity without excess; hematite (iron oxide, $Fe2O3$) gives a red colour.

Observation and the intensity of the colours provide an initial insight into the constitution and functioning of the soil: its fertility, hydromorphy...

B. The texture of a horizon

Granulometry determines *soil texture*. In a soil sample taken from a horizon, different types of compounds are observed: coarse elements (pebbles, cobbles, etc.) larger than 2 mm and fine soil composed of finer elements that are classified into three main groups according to their size:

- clays: particles with a diameter of less than 2p (microns)
- silt: particles between 2 and 50 p (microns) in diameter
- sand: particles with a diameter of between 50 and 2000 p (microns) or 2 mm.

Within a horizon, the proportions of pebbles, sand, clay and silt vary and determine *the texture of the soil*. Depending on the proportion of the different elements, we use the terms: sandy silt, silty sand, clayey sand, sandy clay, silty clay, etc.

Texture influences the soil's water regime (circulation, water retention and capillarity). *Sandy textures* have low water reserves, silty-clay textures are water reservoirs and clay textures can block circulation. These different textures will have an impact on the water supply for vegetation, its development, soil erosion and the supply of groundwater and rivers.

C. The basic organisations of a skyline

The term *elemental organisations* is used in earth science to express the spatial, functional and genetic arrangement of soil constituents. There are three types of elemental organisation: aggregates, voids and pedological features.

- Aggregates

A horizon is made up of small individual volumes called aggregates. Aggregates result from the assembly of elementary particles and the cracking of aggregates. Aggregates define the soil structure.

A horizon can have either a fragmentary or a continuous structure.

In a fragmentary structure, the horizon is made up of aggregates. There are three types of aggregate: rounded, angular and laminated. Within the same horizon, their morphology varies vertically and horizontally. Each type of aggregate has specific meanings in terms of its constituents, genesis, function and fertility. For example :

The rounded, fragmentary structures provide a welcoming (porous, crumbly) environment for roots, allowing gases and water to circulate easily and plant and animal life to thrive.

The angular fragmentary structures indicate the presence of clay, the absence of biological activity and the low organic matter content. The laminated fragmentary structures confirm the significant presence of clay, the lack of porosity and a soil rich in $Ca2+$ and/or $Mg2+$ that cannot be used by roots.

In a continuous structure, there are no aggregates, as in sand.

Figure 4.2 *Aggregates, gravel, sand and clay.*

- Empty spaces

The *voids* (or pores) are the spaces, largely invisible to the naked eye, left free between the particles that make up the horizon, after they have been piled up. Occupied by liquid or gaseous constituents, they serve as circulation channels for water, air and living matter, and characterise the porosity of the horizon. Porosity varies from one horizon to another depending on its morphology, from 20% for a low porosity, very compact horizon to 80% for a very porous horizon.

The morphology and volume of voids depend on a number of factors, such as the size of the soil's constituent particles (clay, sand, etc.), biological activity (roots, animals, earthworms, etc.), the degree to which the soil has dried out, etc. The porosity of the horizons determines the extent of gaseous (carbon dioxide), mineral and liquid exchanges between the soil, the plants and the atmosphere. It plays an important role in the circulation of rainwater and therefore in regulating the water supply to plants and groundwater.

- Pedological features

The term "pedological features" covers nodules, pellicles, bands and pedotubes.

The movement of clay, sand and silt particles, the migration and precipitation of elements in solution (limestone, salts, organic matter), the movement of animals and the development of roots lead to the creation of pedological features, i.e. the formation of accumulations of matter (clay, calcite, carbonates, iron hydroxide, aluminium hydroxide, etc.) in the form of films, nodules and bands.

The presence of this or that pedological feature will determine the characteristics of the soil, its stage of evolution, its fertility, its porosity, etc.

4.3 PLANT LITTER

Generally speaking, litter refers to all dead leaves and decomposing plant debris (primary mineralisation, humification, secondary mineralisation) from epigeous vegetation covering the ground (in forests, gardens, hedge planting, etc.).

Forest litter is mainly composed of softwood leaves and needles, with branches and fruit accounting for only 21% in mixed forests and 20 to 40% in coniferous forests.

Herbaceous plants contribute less than 5% in temperate forests2. The main input of nitrogen (60 to 70%) and certain nutrients such as calcium and phosphorus (75 to 95%) is associated with the annual fall in fresh litter3. Varying on average from 3 (agrosystems) to 10 tonnes (tropical forests) of dry matter per ha per year, the fall of aerial litter is the most important way4 of returning organic matter to the soil5.

In pedology, litter is the superficial layer that covers the soil and forms part of the so-called "holorganic" horizons. It is made up of organic matter; plant residues (leaves, twigs, pollen), fungal residues (spores, mycelium) and animal residues (mainly invertebrate excrement and corpses) that are deposited on the ground throughout the year, still unaltered or only slightly altered.

Generally speaking, litter is only found on the surface of an unworked soil profile, usually in woodland. Pedologists refer to soil layers as horizons: the litter is above and at the origin of the humus horizon.

Plant debris (litter from aerial parts, decomposing roots) constitutes a zone of high biological activity ("microbial hot spot") known as the detritisphere, in the same way as the volume of soil surrounding living roots, the rhizosphere6.

Figure 4.3 *Plant litter (Futura Planete).*

The disintegration of plant litter by animals leads to the formation of *coprogenous humus*: *Moder* or *Mull*; the disintegration of litter by saprophytic soil fungi gives *mycogenic humus* or *Mor*.

4.4 EVOLUTION OF FOREST HUMUS FORMS

Humus forms correspond to a broad definition of the word humus: they encompass all surface horizons containing organic matter. Conditioned mainly by the biological activity of the internal environment, they are closely linked to the composition of the litter and the conditions of the external environment.

The authors have distinguished three basic types of humus that form in aereal environments: Mull, Moder and Mor, and two for hydromorphic environments: peat and anmoor.

In the latter environments, which are more or less temporarily waterlogged (often due to the presence of a water table), humification processes are conditioned by phases of anaerobiosis.

4.4.1 THE MULL

Mulls are forms of humus that are linked to a high level of activity on the part of the soil's fauna and micro-organisms. Their main characteristics can be defined as follows: rapid decomposition of litter, partial incorporation of plant debris within a mixed organo-mineral horizon where "clay-humus" aggregates also form (A_1 or A_h horizon, with a lumpy structure formed by anecitic earthworms, i.e. earthworms that feed on leaves).

The reference mull, described by many authors, is the active eutrophic

mull, formed in conditions favourable to biological activity and on a parent material containing sufficient, but not excessive, amounts of clay and calcium.
It is also characterised by its high A1 nitrogen content. The other forms of mull have a similar morphology, but show notable differences: carbonate mull, thick and containing active calcium carbonate; acidic mull, not very thick, formed on material poor in calcium, but containing free iron; andic mull, thick and dark, with a fine structure, formed on ash or volcanic rocks; finally, chernozemic mull of the forest-steppe in a continental climate, whose Ah horizons, dark grey or black in colour, are exceptionally thick.
Mull is a type of humus that forms in an oxygen environment (aerobiosis). It is brownish in colour and has a granular structure. It develops in a biologically active environment (significant pedofauna) and has a pH close to neutral (ph = 7).

4.4.2 MOR

Mor is a form of humus produced by the slow and incomplete decomposition of plant litter by saprophytic soil fungi; a *mycogenic humus*. They give a more or less thick fibrous organic horizon (Ao or O) superimposed on the mineral horizons.
Humification is practically non-existent, no clay-humus aggregates are formed and the structure of the most transformed horizons (H) is limited to an association of highly divided fresh organic matter and the faecal pellets of lower arthropods.
Mor is a type of humus that forms in an oxygen environment (aerobiosis). It has a fibrous or laminated structure. It forms on siliceous rocks and under thorny vegetation.

4.4.3 MODERATORS

The Moder have intermediate characteristics, although they are closer to those of the Mor than to those of the Mull: a fauna of enchytreids (small whitish or transparent worms) and surface earthworms (epiges), and a structure of the H horizon comparable to that of the Mor, albeit less fine.
Moder is therefore a humus formed by aerobiosis, mainly in forests. Earthworms are fairly rare compared with the Mull. Detritivores are represented by microarthropods such as mites and springtails and by small annelids, the enchytreids. Mulls, like Moders, *are humus coprogenes*.

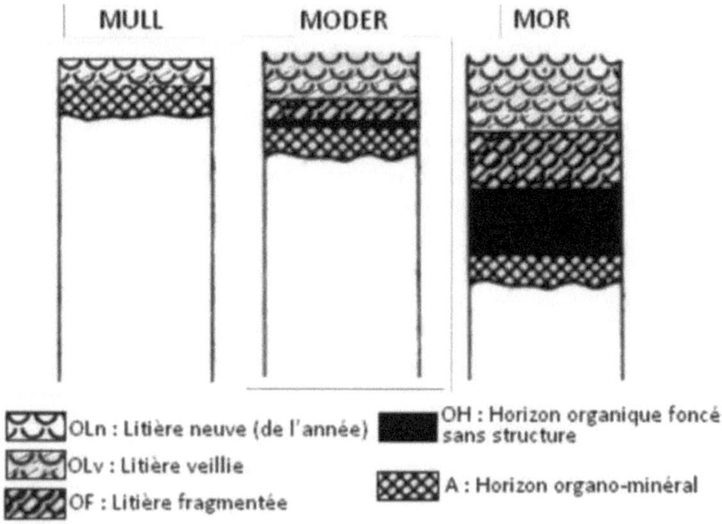

Figure 4.4 *The different types of humus*

4.5 TYPES OF SOIL

There are eight main categories of zonal soils on the earth's surface: *podzolic soils, chernozemic soils, desert soils, ferralitic soils or latosols, Mediterranean red soils, tropical ochre ferruginous soils, tundra soils and mountain soils.*

4.5.1 PDOZOLIC SOILS

A. Podzols

Podzols form in *cold climates* with *high rainfall*; the climax vegetation is the coniferous forest (taiga).

The *litter* is poor and fibrous, and transforms more easily into *mor*, as the climate is unfavourable to the activity of earthworms and bacteria. The mortar formed on the surface by the action of fungi is thick, and mineralisation is slow.

Figure 4.5 *Podzols in the boreal regions*

In *wet* periods, the mor remains stuck in the mineral soil with a large quantity of aggressive *fulvic* (humic) *acids*, resulting in *podzolisation*, decomposing the clay (montmorillite) into silica and aluminium sesquioxide.

Aluminium and iron sesquioxides, complexed by organic acids, are carried to depth where they form highly coloured *B accumulation horizons*.

Below the very thick Ao horizon, the podzolised and leached Ai horizon, composed mainly of silica, takes on an ash-grey hue characteristic of *podzol* (ash).

In periods of *drought* and intense *evaporation*, the accumulated solutions can become supersaturated, forming a hard crust known as *allios or orstein*, which prevents drainage and root penetration.

Worldwide, podzols cover around 15 million km2 in Russia, Scandinavia and Canada.

These soils *are not very* favourable for plant nutrition (very poor base) and are difficult to improve (deficient clay complex). They are best suited to their climax coniferous vegetation, the biosphere's main *reserve* of *cellulose* for paper manufacture.

Podzols are therefore acid soils (pH below 5), characteristic of boreal climates (the climax soil of the taiga) and of certain sites in humid temperate climates (with coniferous or moorland vegetation).

B. Brown forest soils

Brown forest soils are formed in humid temperate climates; their climax vegetation is the deciduous deciduous forest, during its foliage in the cold winter. These are typically *mull* soils, characterised by easy mineralisation of the litter, and consequently by rapid cycling of biogenic elements.

Figure 4.6 *Brown forest soil*

The clay colloids, based on montmorillonite, are well stabilised due to their high exchangeable status; they are mixed with iron oxides which give them their characteristic colour. There is no degradation of the clay, but rather a slight leaching of the clay (*soil leaching is the displacement of nutrients from the soil to deeper layers, caused by precipitation*).

The A B C profile comprises the following horizons:
A_1 black brown mull, a few cm thick;
A_2, brown, with lumps ;
B brown, more clayey, with a polyhedral structure.
Brown soils, or Brunisols, are the classic form of evolved soil found under deciduous forests in temperate zones (Western Europe, Canada and the USA). They generally contain mull-type humus, with a pH of the A horizon of between 5 and 6.

4.5.2 CHERNOZEMIC SOILS

A. Chernozems or black earth

Chernozems form on calcium and magnesium carbonate bedrocks (most often), in a semi-arid continental climate. The climax vegetation is the dense gramineous steppe.

The *profile*, of the *AC type*, is characterised by an A_1 horizon, black and very thick (60150 cm), saturated in calcium and magnesium, the humic colloids are flocculated (the soil is very aerated), and form well aggregated complexes with the clay, rich in grey humic acids with a high nitrogen content.

The abundance of humus is due to: the huge organic mass provided by the Gramineae; biological activity limited to the favourable periods of the semi-arid climate; and the presence of clay, which protects the humus from microbial decomposition.

The highly-developed root system of Gramineae, located in this horizon, enables plants to rapidly reabsorb mineral elements from the decomposition of humus.

Figure 4.7 *The tchernozems of the DRC.*

Chernozems are found in Central Europe, Ukraine, Central Asia, the American prairies and the Argentine Pampas. These are the most fertile soils in the world, and very suitable for growing cereals.

B. Chalky soils and sierozems

These soils are formed in more arid continental climates and are less rich in organic matter and less decarbonate at the surface than chernozems.

The steppe vegetation is sparser. The A1 surface horizon, about fifty centimetres thick, is coloured chocolate brown by calcareous humus.

In drier regions, the soil is low in humus, and the general colour of the soil is brown (*brown* steppe *soils*).

At the edge of deserts, the soil is poor in humus and devoid of iron oxides, with a grey, carbonate-rich surface horizon. Intense evaporation can lead to the efflorescence of a gypsum-limestone crust on the surface (sub-desert *grey soils* or *sierozems*). Sierozems are less fertile than chernozems.

4.5.3 DESERT SOILS

Desertic soils form under vegetation that is naturally very sparse.

The decomposition of the parent rock is purely mechanical; the absence of water does not allow chemical decomposition. There is no humus. A distinction can be made between the *reg* or stone desert and the *erg* or sand desert.

Evaporation brings large quantities of biogenic elements to the surface. These soils are potentially very rich and can be productive if sufficient water can be brought to them.

The large desert areas increasingly include numerous zonal or azonal soils: dirty soils, oasis soils, ouadi soils, etc.

Figure 4.8 *The desert floor (the Wadi Rum desert).*

A desert is an area of land where rainfall is rare and scarce, and where, as a result, living conditions are hostile for plants and animals. Deserts are extreme environments. The lack of vegetation exposes the unprotected surface to the process of denudation.

4.5.4 FERRALITIC SOILS OR LATOSOLS

A. Latosols

In a tropical, humid or equatorial climate, the decomposition of sea-rocks takes place much more vigorously than in a temperate climate; there is a strong tendency for silica to leach out and for aluminium and iron sesquioxides to accumulate in a deep soil with often distinct horizons; biogenic elements leach out to a considerable depth.

If the source rocks are *basic*, the bases and silica are rapidly leached out, leaving a mixture of iron hydroxide and aluminium or bauxite rich in free alumina or *gibbsite*, the main aluminium ore (*ferralites*).

If the parent rocks are *acidic* and rich in silica, some of the silica combines with alumina to form poorly fertile clays of the *kaolinite* group, mixed with iron hydroxide with little free alumina (*ferralitic soils*).

With *or without a* thick, red, iron and alumina-accumulating B horizon, these soils are not very fertile. They are best suited to the climax vegetation of dense equatorial forests.

Figure 4.9 *Ferralitic soil*

Latosols are tropical forest soils with relatively high levels of iron and aluminium oxides. They are generally classified as depleted Oxisols or Ferralsols.

B. Formation of armour

When the forest is destroyed by man, left to fallow and transformed into savannah, the B horizon, enriched in Fe_2O_3, is finally denuded by erosion, and thus exposed to the sun and bush fires; there is dehydration of the iron oxides, which harden into pisolitic concretions (spherules the size of peas) or vermicular concretions.

These concretions fuse together to form spongy carapaces or armour with the consistency and colour of brick.

This is known as laterite (Latin for brick), a term that pedologists now replace with ferralite.

Ferralitic soils can be of several sub-types: red, yellow, ochre or grey.

When this armour is naturally exposed, or when erosion has removed the horizons covering the fossil armour, the soil can be said to be dead; impervious to water and roots.

4.5.5 MEDITERRANEAN RED SOILS

Terra rosa soils are formed in typical Mediterranean climates (wet winters and very dry summers), and on hard limestone bedrocks.

Under the action of heat and high humidity, deep alteration of the bedrock releases clays and quantities of *red* iron oxides.

Seepage water, which dissolves at the surface of the rock, forms a *lighter-coloured horizon* at depth, where carbonates accumulate.

Terra rosa is formed locally, where vegetation is sparse, as a result of erosion of the Mediterranean brown soil. Their formation may also be the result of an ancient evolutionary process, dating back to a geological period with a humid tropical climate (*fossil soils*).

4.5.6 TROPICAL OCHRE FERRUGINOUS SOILS

In dry tropical climates, in the spinisilve zone, or tropical semi-desert steppes, the soils, which are deeper than in temperate regions, are richer in iron oxide, and coloured yellow or ochre.

These soils are rich in ferruginous concretions, but do not form armour; they form the *transition between* Mediterranean soils and ferralites.

4.5.7 TUNDRA SOILS

Tundra soils are characteristic of the sparsely vegetated *polar regions*. Low temperatures reduce chemical and biological activity.

An *AC profile is* formed on undecomposed bedrock or permanently frozen subsoil.

The organic matter decomposes very slowly to form a very thick, *peaty brown humus*. Leaching and podzolisation are low. They form surfaces unsuitable for cultivation.

Figure 4.10 *Tundra soil*

This is a plant formation found in cold, polar or mountain climates, consisting mainly of a single plant layer.

4.5.8 MOUNTAIN SOILS

All *types of soil are* found here, covering all the major *mountain ranges*, the most fertile being brown forest soils, chernozems and alluvial soils.

Figure 4.11 *Mountain soils*

4.6 MAIN CHEMICAL ELEMENTS IN THE SOIL ESSENTIAL FOR PLANT DEVELOPMENT

There are around 50 elements that can make up inorganic substances in plants. But it is important that these are *essential*. An *essential element* is a chemical element that the plant needs during its growth cycle, its reproduction, in short its development cycle, producing another generation.

If we are to obtain information about the nutritional requirements of plants by studying their chemical composition, we need to make a clear distinction between *essential elements* and elements that are only present in the plant.

The minerals present in the soil reflect to some extent the *chemical composition of the soil*. For example, a plant growing in mine tailings may contain gold or silver, but these have no nutritional function.

To determine which chemical elements are essential, researchers are using *hydroponics,* which involves growing plants without soil, directly in mineral solutions or in an inert medium containing these solutions. This technique has enabled us to find 17 elements that are essential for all plants, nine of which are major elements and the others minor elements.

Figure 4.12 *Hydroponics*

The *nine major elements* or *macroelements* are elements that plants need in significant quantities. Six of them are the major constituents of organic substances that form the structure of a plant: carbon, oxygen, hydrogen, nitrogen, sulphur and phosphorus. The other three are calcium, potassium and magnesium.

The other essential elements are called *minor elements* or *micronutrients*, because they are elements that a plant needs in very small quantities: iron, chlorine, copper, manganese, zinc, molybdenum, boron and nickel in particular. They are only useful as *cofactors* in enzymatic reactions.

It should be noted that the levels of different minerals can vary considerably from one plant species to another, or even from one organ to another.

It should also be noted that these minor elements are essential to the life of a plant, and any *deficiency* in one or more of them can weaken or even kill the plant.

The chemical elements essential for plants are absorbed in the form of mineral *ions*, *water* or *carbon dioxide*. In general, mineral ions are absorbed by the active transport mechanism at the level of the absorbing hairs or mycorrhizae, water is absorbed by osmosis and carbon dioxide by simple diffusion at the level of the stomata.

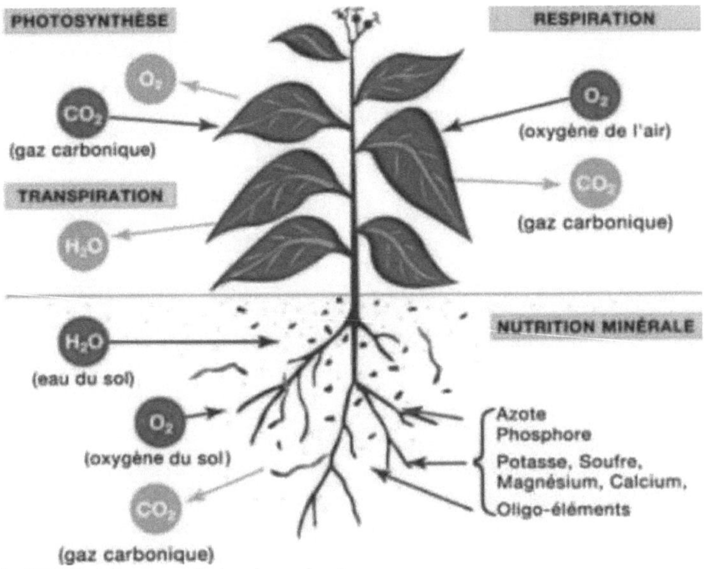

Figure 4.13 *Nutrient uptake by plants*

4.7 ABIOTIC FACTORS IN THE SOIL

Soil is a complex environment inhabited by an abundant and highly diverse fauna. There are many abiotic factors involved in soil ecology. We will limit ourselves to the main ones.

4.7.1 WATER IN THE SOIL

Water in the soil is found in specific states, which pedologists have classified into four categories.

A. Hygroscopic water

It comes from atmospheric humidity and forms a thin film around soil particles. It is retained very energetically and cannot be used by animals or plants.

B. Non-absorbable capillary water

It occupies pores smaller than 0.2^ and 8ц in diameter. It is normally absorbed by plants outside rainy periods and allows the activity of bacteria and small protozoa such as flagellates, or thecamoeubians in certain cases.

C. Gravity water

It temporarily occupies the largest pores in the soil. This water runs off under the action of gravity, except when drainage is impossible. A distinction is made between fast-flowing water, which circulates in the large diameter pores and exists only in sandy soils, and slow-flowing water, which can remain for several days in the finer, non-capillary pores.

To assess the amount of water available in a soil for vegetation, we determine the permanent wilting point, i.e. the amount of water (in % of total weight) that is still in the soil when the plants begin to wilt permanently. This point can be determined either by a biological method (a plant is grown in a pot and then left to wither) or by a biological method (a plant is grown in a pot and then left to wither).

4.7.2 SOIL TEXTURE AND STRUCTURE.

A. Soil texture

Soil texture corresponds to its granulometric composition. The various elements making up the soil are classified in the following categories:

- Coarse sands from 0.2mm to 2mm
- Fine sands from 0.2mm to 20^.
- Mineral colloids (mainly clays) of less than 2 д elements of more than 2mm make up gravel.

B. Soil structure

Soil structure is another concept. When the colloids in the soil are dispersed and their grains isolated, whatever their size, *the structure is said to be particulate*. When the colloids are flocculated into more or less stable aggregates, *the structure is said to be lumpy*. Flocculation occurs when *Ca++* and *Mg++* ions are abundant. Soil structure is very important. It is particularly important for soil aeration. Soils with a lumpy structure are permeable and well aerated, while soils with a particulate structure are more or less impermeable, depending on their colloid content, and therefore more or less poorly aerated.

Soil *porosity* depends on both the *texture* and *structure of* the soil.

Granulometry plays a major role in the distribution of animals in groundwater and beach sands.

4.7.3 SOIL AERATION

Soil porosity regulates the circulation of water, air and many animals. A compact, low-porosity soil can prevent temperature- and moisture-sensitive animals from migrating vertically, thus making their existence impossible.

4.7.4 SALINITY

Dirty soils or halomorphic soils come in a variety of forms. Some are saline soils or *solontchaks*. Their pH is below 8 and sodium forms no more than 50% of the cations in solution. Others are alkaline soils, or solonetz, containing an excess of sodium that forms sodium carbonate, with a pH of up to 9. Dirty soils have very specific flora and fauna.

4.7.5 pH

It cannot be said that soil pH acts directly. It is in fact a result of all the various pedological factors. The pH is essentially dependent on clay and humic colloids. These are electronegative and surrounded by H^+ , Ca^{++} , Mg^{++} , Na^+ and K^+ cations. The pH also depends on the nature of the plant cover and climatic conditions (temperature and rainfall). The pH of calcareous soils is around 8; that of dirty soils can reach 9.5; in sphagnum peat and podzols the pH can be below 4.

Protozoa can withstand variations in pH from 3.9 to 9.7 depending on the species; *Paramecium* can withstand pH from 5 to 9.

Lombricides can withstand pH values of 4.4 and above; between 4.5 and 8, preferences are difficult to establish due to the interaction with various factors such as calcium content.

Molluscs depend on the pH of the soil because of their need for limestone, at least those species that have a shell. The greatest number of species are found in soils with a pH of 7 or slightly higher (7 to 8).

4.8 IMPORTANCE OF SOIL IN PLANT NUTRITION

They allow water to infiltrate, contribute to water storage and store plant nutrients. They resist compaction and provide a healthy, airy root environment.

Soil and vegetation are therefore interdependent. Fertile soil encourages plant growth because it provides plants with nutrients, acts as a water reservoir and supports the plants that take root in it.

REVISION QUESTIONS ON CHAPTER 4

1. Explain these terms: soil, bedrock, pedology, edaphology, edaphic fauna, an abiotic feature, a biotic feature, flora, a natural disaster, a soil horizon, leaching, podzolisation, humus, plant litter, zonal soils, interzonal soils, azonal soils, soil texture, soil structure, soil porosity, soil profile, soil cover, topography, tundra, mangrove.
2. What are the main stages in soil formation?
3. Name the main soil horizons and give the essential characteristics of each.
4. Explain how soil horizons are formed.
5. What is the essential difference between mycogenic humus and coprogenic humus? What do Mull, Moder and Mor mean?
6. What are the 8 main categories of zonal soils that cover the surface of the globe?
7. What are the main characteristics of the following soils: podzols, latosols, chernozems and solontchaks?
8. Describe these terms for plants: essential chemical elements, major chemical elements, minor chemical elements, hydroponics, deficiency disease, chemical elements that are co-factors in enzymatic reactions, stomata, mycorrhizae, mutualism, symbiosis, osmosis, active transport, absorbent hairs.
9. Show that the absorbent hair is a cell adapted to its function. And what is an adaptation?
10. Show how edaphics can be responsible for the repartition and distribution of living organisms.
11. Name 5 important soil abiotic factors.
12. How important is soil in plant nutrition?

CHAPTER 5

LES ROCHES

5.1 MINERAL

A *mineral* is a naturally occurring inorganic chemical species, usually in the form of a crystalline solid. Classification is based on their chemical and crystallographic characteristics.

Around 98% of the Earth's crust is made up of eight chemical elements: O, Si, Al, Fe, Mg, Ca, Na and K. These elements form the basis of the minerals that make up the Earth's crust.

Here are a few examples of the main classes of minerals: silicates (which alone make up more than 98% of the Earth's crust), carbonates, phosphates and sulphates.

The main *silicates* are quartz, feldspars, amphiboles, pyroxenes, micas, biotites and olivines.

The physical properties of a mineral depend on its colour, cleavage or fracture, hardness, luster and density.

Other *physical properties* of a mineral such as transparency, tenacity, taste, touch, odour, fusibility, radioactivity, magnetism (magnetite), conductivity, etc. can be used either to identify a particular species or for industrial purposes.

5.2 LES ROCHES

Rocks are materials formed, in general, from an assembly of natural minerals. A rock is therefore an assembly of minerals, whereas a mineral is an ordered assembly (crystal) of atoms.

There are 3 main groups of rocks, distinguished by the way they are formed: magmatic rocks (or igneous rocks): resulting from the crystallisation of a magma, sedimentary rocks: these are exogenous rocks, i.e. formed on the Earth's surface, and metamorphic rocks: rocks resulting from the transformation of an original rock into a solid state as a result of an increase in pressure and/or temperature. During this transformation, new minerals (neoforms) generally crystallise out and the rock acquires a particular texture (e.g. schistosite = layered debit).

All these rocks, as soon as they come into contact with the atmosphere, whatever their origin, composition or texture, are subject to the action of various factors that lead to their *alteration*.

Subjected to physical, chemical and biological *factors*, superficial continental rocks, even the most resistant, are altered. Climatic conditions have a major influence on how this alteration takes place.

The *observation* of rocks that have undergone alteration in a given geological period can therefore provide information about the conditions that prevailed at that time.

Torrents, rivers and streams provide good observation sites for the *erosion*, *transport* and *removal of* material by stream water. *Hjulstrom*'s

experiment established the importance of material size in the *erosion phenomenon*.
sedimentation.

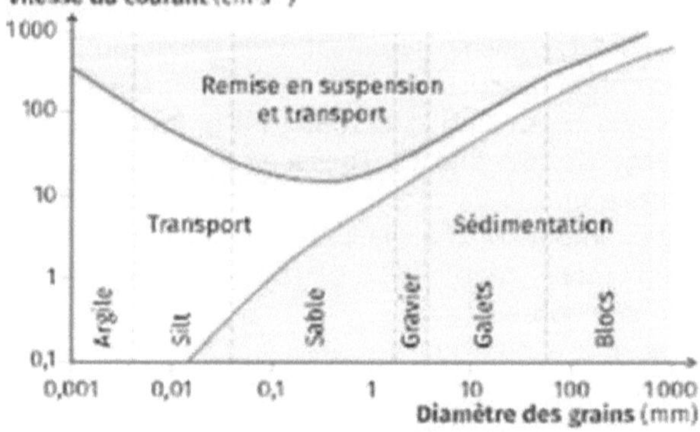

Figure 5.1 *The Hjulstrom diagram*

The Hjulstrom diagram shows the activity of a detrital particle as a function of the speed of the current in the water (sea, river) in which it is found, and as a function of its size. It was invented by the Swedish geographer Filip Hjulstrom (1902-1982).

Filip Hjulstrom
(1902-1982).

For example, a grain of sand 0.3 mm in diameter can be torn from the "substrate" if it is attached to the bottom when the current speed reaches 100 cm/s. If the current speed is 10 cm/s, circulating sand grains of this size will continue their journey; and at a low speed of around 1 cm/s, the grains will settle.

We can therefore define the activity of the particles, i.e. whether they are eroded, transported or deposited, as a function of their size and the speed of the current.

This diagram dates from the early twentieth century and is now only of historical value, although its simplicity remains attractive. The diagram's shortcomings include the fact that it does not take account of water depth

and, more seriously, that it does not show that sedimentation is caused by a deceleration of the flow and erosion by an acceleration of the flow.
The *granulometry* of a sediment studies not only the classification of grains according to size, but also their shape, appearance and traces of wear. It provides a wealth of information about the environment and sedimentation conditions.
The materials resulting from the degradation of rocks are transported either in the form of *fragments* of varying sizes, or in the form of *ions* in solution. After numerous episodes of deposition and recovery, they finally end up in *the sea*, where they accumulate as *sediments*.

5.3 MAGMATIC ROCKS

5.3.1 Internal structure and composition of the globe

For a long time it was believed, and this idea is still held by the general public, that the Earth's interior was the seat of a "central fire" and that it was made up of molten matter that rose to the surface during volcanic eruptions.
We have known for a very long time that temperature increases with depth, and miners were well aware of this when they went down to the bottom of the mine. The increase in temperature measured at the surface is of the order of 3°C/100m.
Quick calculation: 3°C/100m = 3000°C at a depth of 100km. Now 3000°C is well above 1200°C, the average temperature of a basaltic lava in Hawaii. It's tempting to say that the Earth's interior is molten. Well no, because we're forgetting one thing: pressure inhibits fusion (solid/liquid transition).
Today, thanks to geophysics, we know very well that the vast majority of the globe is solid, with a few exceptions.

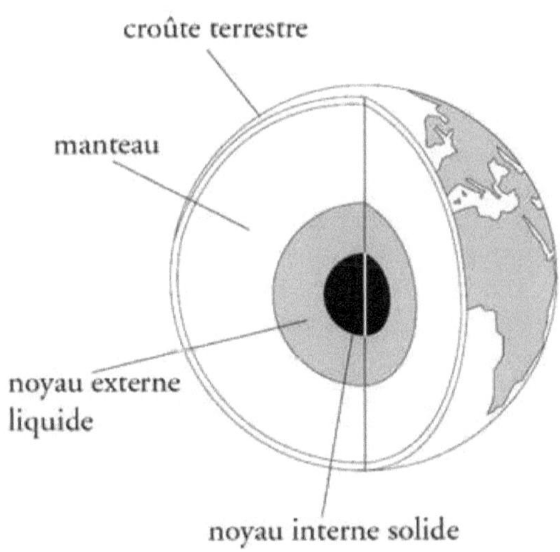

Figure 5.2 *Representation of the earth's crust*

Earth's radius: 6378 km at the equator, 6356 km at the poles. From 0 to 30 km depth: the Earth's crust. At 30 km, we have what is known as the Mohorovicic discontinuity (or MOHO). Density of the crust: 2.7 to 2.9.
From 30 to 670 km: upper mantle and from 670 km to 2900 km: lower mantle.
The mantle is essentially composed of peridotite (a rock made up of olivine, pyroxenes, etc.). The boundary between the upper and lower mantle is determined by a change in the crystalline phase of olivine (the peridotite mineral). We will look at this again in the seismology course.
The earth's mantle and crust are solid.
From 2900 km to 5000 km: outer core, composed of Fe (86%), S (12%) and Ni (2%). The outer core is liquid (this will also be covered in the seismology course). From 5000 km to 6356 km: inner core, composed of Fe (80%) and Ni (20%).
There is another breakdown of the surface envelopes: from 0 to 100 km: Lithosphere and from 100 to 670 km: Asthenosphere.
One thing to remember: SiO_2 (generally known as silica) is the most abundant molecule in the Earth's crust and mantle. SiO_2 is the basis of the most widespread family of minerals on Earth, the silicates (quartz, micas, feldspars, amphiboles, pyroxenes etc. are all silicates). Quartz is pure SiO_2.
In conclusion, the Earth is essentially solid and formed from rocks made up of the main minerals: silicates.
The birth of a magma in this solid Earth is therefore a rather exceptional phenomenon that requires very specific conditions, in particular a local or regional thermal imbalance. Magmas are generally formed at depths of between 10 and 200 km.
5.3.2 Origins of magmas
A magma is a natural molten bath of silicate or aluminosilicate nature. The SiO_2 (silica) content varies from 40 to 75% by weight.
Magma is made up of 3 phases:
A liquid phase which represents between 10 and 70% of the magma,
A solid phase: magmas result from the partial melting of a rock. Pieces of the primary rock therefore remain, as do pieces of rock that have been incorporated into the magma as it rises through magma conduits.
A gaseous phase (mostly water), in variable proportions (1 to 7%)
Magmas are also characterised by their temperature, which varies from 700°C for granitic magma to 1200°C for basaltic magma. Magmatism in the context of plate tectonics.
Magmatic phenomena occur in 3 main types of zone on a plate scale:
Divergent plate boundaries (oceanic ridges),

Convergence limits,
Intraplate sites,
There are 3 extreme possibilities for partially melting a peridotite in the mantle (to take it above solidus):
Case 1: a drop in pressure can lead to the formation of magma if the temperature remains more or less constant. This happens at ridges. At this level, the lithosphere stretches and thins, causing a drop in pressure in the asthenospheric mantle below, which tends to rise (see poly diagram). The mantle rises without losing much heat (adiabatic decompression), which causes the magma to melt and form. Once formed, this magma rises to the surface and gradually forms the oceanic crust (made up of basalt).
Case 2: Increase in temperature: In the middle of the plates, there are volcanoes which are usually hot spots (e.g. Hawaii). They are located above very slow ascending asthenospheric currents (don't forget that the mantle is solid, but it deforms (see lesson 2)) that animate the mantle. The heat supplied by these rising hot plumes is enough to increase the temperature of the base of the lithosphere and cause partial melting of the mantle (between 70 and 170 km below the surface).
Case 3: Lowering of the melting point by the addition of water: When water is added to the mantle, the solidus curve of the peridotite is lowered. This is what happens in subduction zones. In these zones, the oceanic crust (which has generally remained under the water of the oceans for several tens of millions of years) plunges into the Earth's mantle. It carries with it a large quantity of water embedded in hydrate minerals such as clays, micas, amphiboles etc... As the plate plunges, the pressure increases and the minerals gradually destabilise and lose their water. This water, released at a depth of around a hundred km, lowers the melting temperature of the mantle and melting occurs. All the volcanoes around the Pacific (belt of fire) are evidence of this type of magmatism.

5.3.3 Partial melting and fractional crystallisation at the origin of the diversity of magmatic rocks

When we compare the chemical compositions of mantle source rocks, which melt to form magma, and magmatic rocks, which are formed from the crystallisation of this magma, we see that they are very different. Why this difference? There are several mechanisms behind this, and 2 of them are extremely important: partial melting and fractional crystallisation.

A. Partial merger :

When a rocky material melts, it very rarely melts completely (except sometimes when the continental crust melts under special conditions). In most cases, melting is only partial and rarely exceeds 30%. This partial melting is incongruous. In other words, the liquid obtained does not have the same composition as the original rock. How is this possible? Well, a rock is an association of minerals of different natures. Not all the minerals melt at the same temperature, and the chemical elements behave differently during fusion. Some elements are said to be "hygromagmatophilic", meaning that they prefer to pass into the fluid phase.

For example: potassium K passes very quickly through the liquid phase. So :

If the melting rate is low, K and other alkalis will pass rapidly into the fluid phase. The liquid will therefore be rich in K and alkalis.

If the melting rate is high, all the K and alkalis will be in the liquid, but other elements that are less hygromagmatophilic will also pass into the liquid. The magma will therefore be proportionally less rich in K and alkalis (because they will be diluted, as it were, with the other elements).

Therefore : - if the melting rate is low (5%): liquid rich in alkalis and alkaline earths. We speak of alkaline magma.

If the melting rate is high (30%): alkaline-poor liquid. We speak of tholeiitic magma.

B. Fractional crystallisation

The difference in the behaviour of the chemical elements observed during partial melting is also observed when the magma crystallises as it cools.

As magma rises to the surface, it can remain in magma chambers for varying lengths of time, where it gradually cools and begins to crystallise. The first minerals to form are ferromagnesian (Fe+Mg). So the magma fluid will gradually become depleted in Fe and Mg! and conversely, the fluid will gradually become enriched in Si and Al.

So an initial magma rich in Fe and Mg will gradually become depleted in these 2 elements and enriched in Si and Al as the magma cools and fractionally crystallises.

This is known as "magmatic differentiation".

Rocks and magmas are generally classified according to their Si content:

Acidic rocks (magma) = $SiO_2 > 65\%$ (example: granite), (the magma has undergone strong differentiation)

Intermediate rock = $52\% < SiO_2 < 65$

Basic rocks = $45\% < SiO_2 < 52\%$: (example: basalt)

Ultrabasic rocks = $SiO_2 < 45\%$ (e.g. peridotite in the mantle)

Done, granite results from a very different magma, far removed in terms of composition from a magma of mantle origin. As the magma has undergone strong differentiation, it has cooled, so the temperature of a granitic magma will not be very high (700°C).

Basalts, on the other hand, are basic rocks, and therefore quite 'poor' in SiO_2. The original magma has undergone little differentiation and little cooling. Basaltic lavas are much hotter (1200°C), which is why they are very fluid and why we see such beautiful fluid flows in Hawaii.

5.3.4 Deposit types for magmatic rocks

A gabbro and a basalt are 2 magmatic rocks that have more or less the same chemical composition. However, in gabbro, you can see millimetre-sized crystals, whereas in basalt, you can hardly see any naked crystals (just a few). Why this difference?

From the same magma, the texture (presence or absence of minerals visible to the naked eye) of the rock obtained depends on the rate of cooling. The slower the cooling, the more minerals will grow. If, on the other hand, the cooling rate is high, the minerals will not have time to form and will be microscopic. If the cooling rate is extreme (when magma meets water), no crystalline minerals appear, and the rock takes on a so-called "vitreous" structure (e.g. obsidian). This is the quenching phenomenon.

We will distinguish between :

Rocks with a grainy texture where all the minerals are visible to the naked eye and are millimetre-sized,

Rocks with a micro-grained texture, where a few minerals are visible to the naked eye but the bulk of the rock is made up of minerals that are perfectly visible under the microscope.

Rocks with a vitreous texture, where there are a few crystals but the bulk of the rock is in the form of glass (a non-ordered structure of atoms, unlike crystals).

What does the rate of cooling of a magma depend on? The temperature contrast between the magma and its casing (what's around it). So it depends on where the magma crystallises.

If the magma cools at depth (where the ambient temperature is high enough), cooling will be slow and the minerals will grow. The resulting rock will be a "plutonic" magmatic rock with a grainy texture (e.g. granite, gabbro).

If the magma arrives at or very close to the surface, it cools much more quickly because the ambient temperature is lower. The resulting rock is a "volcanic" magmatic rock, generally micro-grained or vitreous in texture.

5.4 SEDIMENTARY ROCKS

These are exogenous rocks, i.e. formed on the Earth's surface, representing 5% of the crust by volume and covering 75% of its surface. They are highly variable because their genesis depends on a number of factors: the initial nature of the damaged and altered materials, the type of alteration, the mode of transport, the mode of deposition and the methods of diagenesis; all the processes that transform a loose sediment into a compact, hard rock.

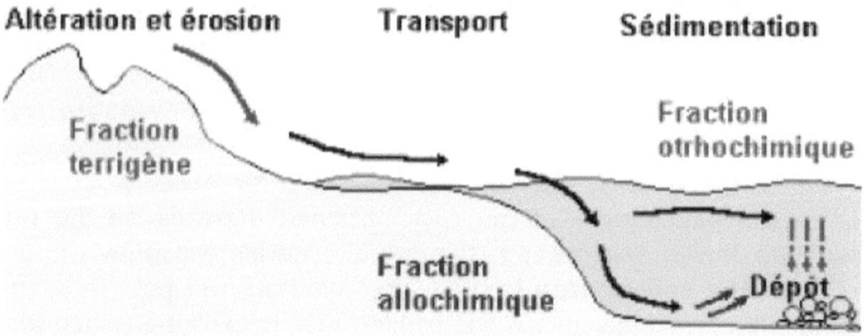

Fig. 5.3 *Sedimentation*

5.4.1 Origin of sedimentary components

The particles that form sedimentary rocks are the result of physical or chemical attack on continental formations. This attack is known as alteration.

The ability of a rock to alter (alterability) depends on a large number of factors: nature and composition of the rock, composition of the water, temperature, degree of fracturing of the rock, etc.

5.4.2 Physical (mechanical) alteration

Mechanical weathering of rocks by physical alteration predominates in regions marked by significant temperature variations. Alternating freezing and thawing causes rocks to fragment. Physical alteration can be linked to the mechanical action of biological organisms: root growth, lichens, mosses, bacteria, etc.

5.4.3 Chemical alteration

Its action is much more widespread than the physical alteration it often accompanies. Rocks that come to the surface, such as granite, are often formed from minerals that have crystallised at depth, i.e. at high temperature and high pressure. These rocks have been brought to the surface by erosion. The minerals are therefore no longer stable at ambient pressure and temperature. They will therefore tend to destabilise to form new minerals, which will be in equilibrium under surface conditions. The rocks will thus alternate to form soils.

The essential phenomenon of chemical alteration is hydrolysis, i.e. the progressive subtraction of ions from minerals by water.

It is clear that the solubility of minerals in water varies according to the chemical nature of the minerals and the properties of the water (t°C, pH, etc.).

Hydrolysis can be described as follows:

Primary mineral (rock) + etching solution (weakly ion-loaded water) -->> secondary mineral (which will form soils) + leaching solution (ion-loaded water). Remember: quartz is the most stable of all silicates.

5.4.3 Main types of sedimentary rock

A small proportion of the alteration products may remain in situ and form residual deposits (laterites). These include elements that are not easily transported by water, such as aluminium -> Al deposits = laterites.
Most of the products of surface alteration are transported by wind, water, glaciers, etc.... to the sunken parts of the continents and then to the seas and oceans.
The accumulation of debris on the bottom of the sea, rivers or lakes by sedimentation, transported in solid form in the water, gives rise to detrital deposits (also known as clastic or terrigenous deposits) and then to detrital rocks after diagenesis (e.g. sand (rock) = compact, consolidated sand).
The material brought into solution during chemical alteration is transported by the water system to lakes, seas and oceans where it can precipitate through various processes:
- directly by physico-chemical processes (evaporation of water and supersaturation of ions leading to precipitation). The associated deposits are referred to as chemical deposits.
These are sedimentary rocks of chemical origin and often evaporitic rocks.
- thanks to the indirect intervention of plant organisms which, by fixing the CO_2, cause the precipitation of carbonate ($CaCO_3$) ($2HCO_3^- + Ca^{2+} \leftrightarrow CaCO_3 + CO_2 + H_2O$). This gives rise to certain lacustrine limestones.
- substances in solution can finally be fixed in the skeletons of living organisms. When these organisms die, the shells or skeletons accumulate on the seabed and end up forming what are known as "organo-detrital" deposits. This is how most limestone is formed (shell limestone, chalk, etc.). These sedimentary rocks are known as biochemical.
During sedimentation processes, organic matter (animal or vegetable) can be preserved within the deposits and evolve to form peat, coal and all hydrocarbons (oil, gas).
Sedimentary rocks are dominated by the detrital fraction: 86% of sedimentary rocks are detrital, 14% are limestone and 1% are all other rocks.

5.4.4 The sediment cycle

In geology, a ***sedimentary cycle*** is a repetition of sequences of sedimentary rocks. It often corresponds to the period between a transgression and the marine regression that follows it.
The stages in the cycle are: ***alteration*** of a rock at the surface by meteoric agents, erosion, ***transport of*** the products of alteration, ***sedimentation***, which results in the formation of a sediment, ***diagenesis***, which transforms the sediment into sedimentary rock, and finally ***exhumation of*** the rock at the surface, and the cycle can begin again.

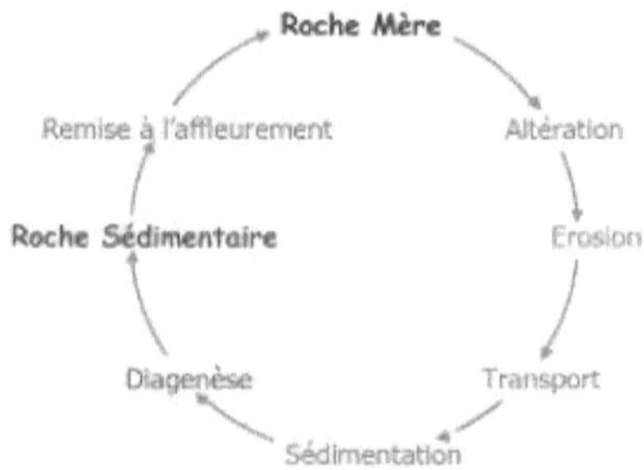

Figure 5.4 *The sediment cycle*

A distinction is made between **marine** sedimentation (coastal, oceanic, bathyal, abyssal, etc.), **lagoon** sedimentation (in areas separated from the sea by a barrier beach) and **continental** sedimentation (wind, river, delta, lake, glacier, etc.).

5.5 METAMORPHIC ROCKS

Metamorphism is the transformation of a rock into a solid state as a result of an increase in temperature and/or pressure, often involving the crystallisation of new minerals, known as neoforms, and the acquisition of specific textures (e.g. schistosite).

The starting rock can be any rock (magmatic, sedimentary or metamorphic).

Metamorphic rocks are formed in areas of intense deformation of the Earth's crust: subduction zones, mountain ranges, etc.

Example:

Granite ^increase in T and P -> limestone gneiss->increase in T and P -> marble

REVIEW QUESTIONS ON CHAPTER 5

1. What are the 8 main chemical elements that make up the Earth's crust?
2. Explain these terms: rock, mineral, sediment, crystallography,
3. Name the 8 types of silicate that make up the Earth's crust.
4. Name the physical properties that are used to identify a mineral.
5. What does granulometry involve?
6. Name the 3 main groups of rocks according to how they were formed.
7. Explain the Hjulotrom diagram using a diagram.
8. Name the 4 categories of magma according to their silicon content.
9. Describe the stages of sedimentation.
10. What is a metamorphic rock? Give two examples.
11. What is a sediment cycle? And describe the different stages in a sediment cycle.
12. Explain the different origins of sediments.

CHAPTER 6

GEOLOGICAL SCALES

6.1 THE GEOLOGICAL CALENDAR

A geological time *scale* is a chronological classification system used, particularly in geology, to date events in the Earth's history.

On *a large scale*, three geological phenomena show the systematic changes that are essential for establishing a chronology of geological time: the evolution of life, the radioactive decay of unstable isotopes and the paleomagnetism of rocks and minerals.

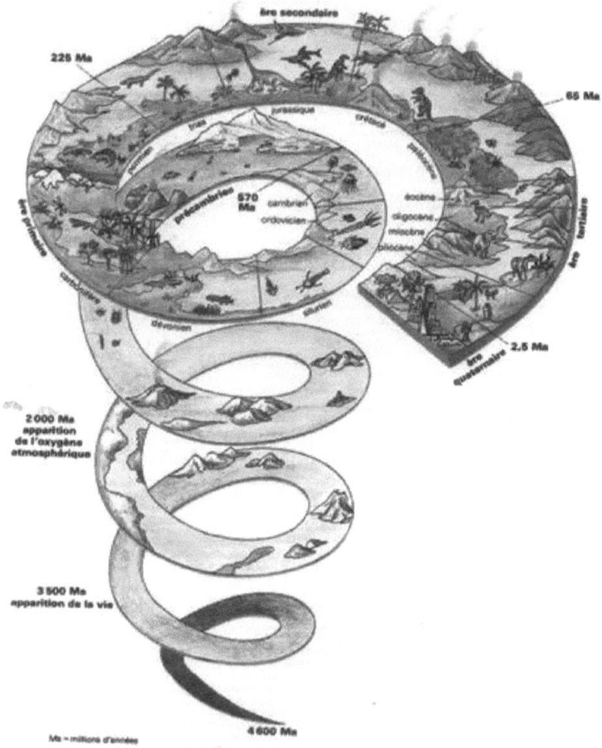

Figure 6.1 *The geological calendar*

Albert ALOMBA, Introduction a la Geologie.

6.2 STRATIGRAPHY

Animal fossils and ancient flora make it possible to classify the terrestrial layers that contain them chronologically; this is known as stratigraphy.

The Earth was formed at the same time as the solar system as a whole, 5 or 6 billion years ago, the solid crust around 4 billion years ago and the first living organisms, which are prokaryotes, i.e. blue-green algae (cyanophycees) and bacteria, appeared around 3 billion years ago. These earliest known fossils are microfossils.

Microfossils are fossilised organisms of small size (mainly less than a millimetre), which can only be observed using suitable tools such as electron microscopes. They differ from macrofossils, which can be observed with the naked eye.

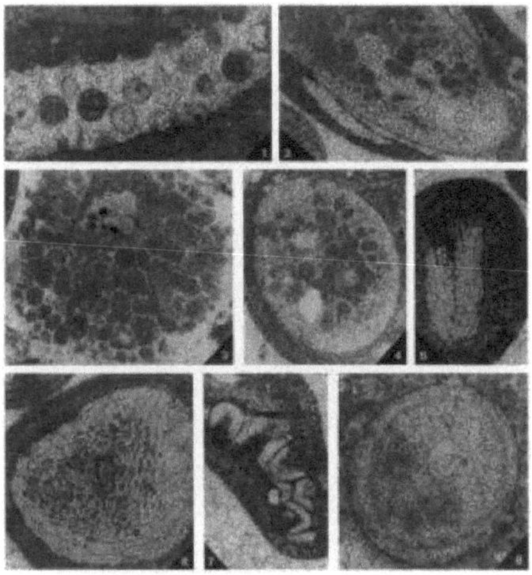

Figure 6.2 *Microfossils*

6.3 DIVISIONS OF GEOLOGICAL TIME

Geological times are the names given to different periods in the history of our planet, which began more than 4.5 billion years ago. These periods make up a scale, known as the geological time scale.

Table 6.1 *Geological time*

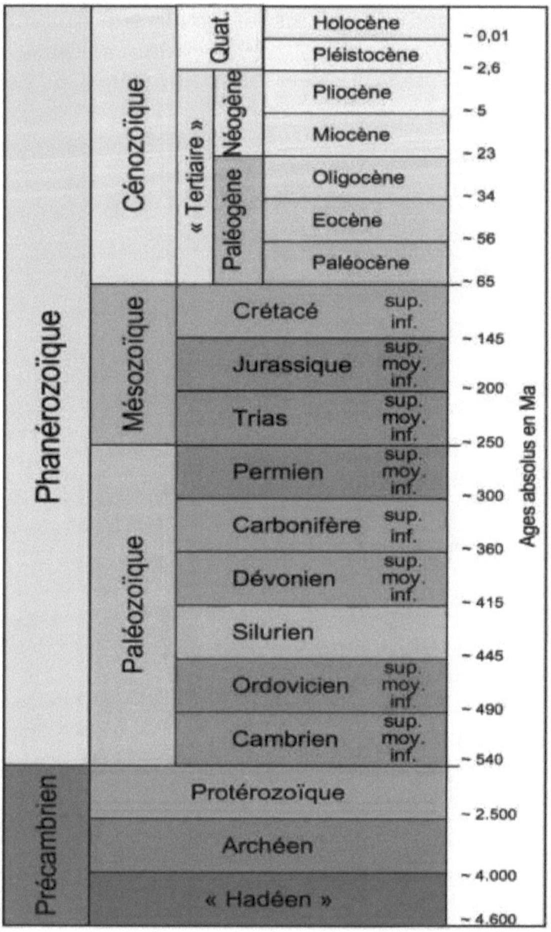

This scale is the result of the work of many scientists who are working, in particular, to date fossils found on Earth.

Traditionally, the Earth's history was divided into four eras, called the Primary, Secondary, Tertiary and Quaternary eras. The time scale used today is a little more complicated, if only because of the words used. But it is more useful for scientific purposes.

The history of the Earth is divided into eons, themselves divided into eras, divided into periods, divided into epochs, divided into stages.

Here are the current subdivisions of geological time:

Aeon: there are four, the Hadeen (creation of the Earth and oceans), the Archeen (appearance of simple Life), the Proterozoic, and the Phanerozoic, which begins in the Cambrian and continues until the present day. The Hadeen, Archean and Proterozoic eras make up the Precambrian, while the Primary, Secondary, Tertiary and Quaternary eras form the

Phanerozoic (see Table 6.1).

Era: there are ten. Seven are in the Precambrian, three in the Phanerozoic, and the last, our era, the Cenozoic. We no longer use the old era names (used in the nineteenth and twentieth centuries) called the Primary Era (the Paleozoic), the Secondary Era (the Mesozoic, the era of the dinosaurs), the Tertiary Era (the Cenozoic), and the Quaternary Era (which has become an integral part of the Cenozoic).

Period: for example, our own: the Quaternary.

Era: ours is the Holocene.

Floor.

Table 6.2 *Chronology of the living world*

ERES GEOLOGICAL	PERIODS	AGES/ 10^6 YEARS	LINES ANIMALS	LIIGNEES PLANTS
Quaternary or ANTHROPOZOIC	Holocene Pleistocene	3,3	Men	
Tertiary or CENOZOIC	Pliocene Miocene Oligocene Eocene Paleocene	65	Current mammals	
Secondary or MESOZOIC	Cretaceous	136		Angiosperms
	Jurassic	150	Primitive birds	
	Triassic	225	Archaic mammals	
Primary or PALEOZOIC	Permian	280		Gymnosperms
	Carboniferous	345	First insect reptiles	
	Devonian	395	Amphibians	Fougeres Mushrooms
	Silurian	430	Archaic poisons	Foams
	Ordovician	500	Agnathes	
	Cambrian	570	Prevertebres	
PRECAMBRIEN			Crustaceans Molluscs Echinoderms Annelids Crelenteres Sponge Protozoa	Eukaryotic algae

In this table, the Quaternary is considered as a geological era in its own right, and the Pleistocene and Holocene as geological periods of the Quaternary or Anthropozoic Eras.

6.3 THE BONOBO OR CHIMPAZE PYGMEE

The Bonobo, Dwarf Chimpanzee or Pygmy Chimpanzee is a species of primate in the Hominid family. Similar to the common Chimpanzee, it differs mainly in its social organisation, which relies on sexual relations and a scapegoat as a means of resolving conflicts within the group.

Figure 6.3 *The Bonobo or Pygmy chimpanzee*

6.3.1 Systematic position of Bonobo

The Bonobo, Dwarf Chimpanzee or Pygmy Chimpanzee (*Pan paniscus*) is a species of primate in the Hominid family. It is closely related to the common Chimpanzee or *Pan troglodytes*.

Regne :	Animalia
Branch :	Chordata
Class :	Mammalia
Orders :	Primates
Family :	Hominidae
Genre :	Pan
Species :	*Pan paniscu*

6.3.2 Some characteristics

With no tail, the bonobo has a slender body, long, powerful arms, a short neck and large, round ears. Its flat face features two bulging eyes, deep nostrils and prominent pink lips.

Life expectancy: 40 years (In captivity)

Gestation period: 240 days

Size: 1.2 m (Adult)

Conservation status : Endangered (Dwindling) Encyclopedie de la Vie

Weight: 34 - 60 kg (Male, Adult), 30 kg (Female, Adult)

Length: 70 - 83 cm (On all fours, from nose to tail).

The female bonobo likes to...show off her bum in order to be seen, while the female bonobo has no problem revealing her private parts.

Bonobo DNA is 98.7% identical to human DNA.

They are adept at "convivial sex", the aim of which is above all social. These couplings are intended to resolve conflicts, for example by appeasing an aggressive individual or if a rivalry breaks out between two members of the community.

While the common chimpanzee can be found in a large number of African countries, this is not the case for the bonobo, which lives exclusively in the Democratic Republic of Congo. Where the species is endemic.

The very name "bonobo" derives from the town of Bolobo on the banks of the Congo River. It was here that the first individuals were observed and captured in 1920.

Over the last 30 years, deforestation has considerably reduced their habitat and the bonobo is an endangered species. There are now just 20,000 bonobos in the wild.

The bonobo community, a matriarchal society Unlike its cousin the chimpanzee.

The females are also very supportive and spend a lot of time together. They have all the power and don't hesitate to join forces to impose their rules and correct the males if necessary.

6.4 THE SPECIAL CHARACTERISTICS OF MAN

Man's anatomy and physiology are close to those of the anthropoid primates: Orangutan, Chimpanzee and Gorilla. But he differs from them in many ways which it will be impossible to list here, for example:

- A perfect upright position.
- A face-lift.
- Three times the cranial capacity of today's anthropoids.
- Extreme mobility in the hand, with a thumb that can be set against the other fingers...

Man differs from anthropoid primates in many other psychological and sociological ways. Primarily the ability to *create tools*. These tools always accompany human remains and enable us to assess the state of *hominisation*.

6.5 THE PRE-HUMANS

Man is not descended from the ape, anthropoids and humans are separate distinct branches. The general opinion of specialists is that the origin of the line that led to man is very old, dating back to the second half of the Tertiary Era, around 20 to 10 million years ago.

Of course, we don't know what tertiary man was, but the great tertiary primates are much more similar to man than to today's anthropoids. This large primate, called the Oreopithecus, had humanoid characteristics (dentition, pelvis, biped...) but was not a human being,

pelvis, biped...) but it probably only represents a small branch parallel to the human branch.

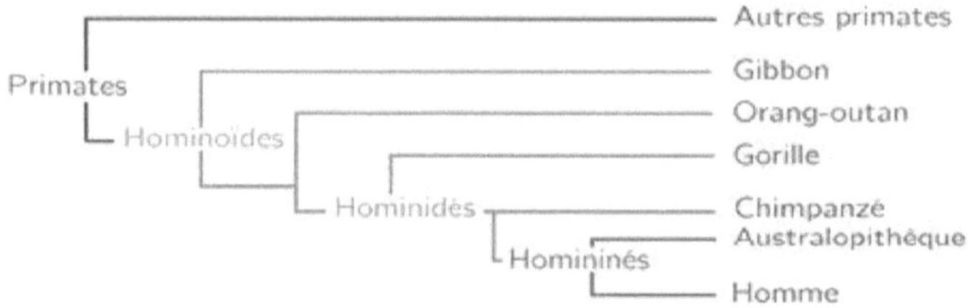

Other primates
Gibbon Orang-utan Gorilla
Chimpanzee Australopithecus
Men

Figure 6.4 *The evolution of Primate paleontology*

6.6 STAGES OF HOMINISATION

A. Australopithecus and *Homo habilis*

The oldest fossils belonging to the humanoid line were found on the eastern edge of Africa, first in the south, hence the name Australopithecus, then more recently in Kenya, Tanzania and Ethiopia.

Taken as a whole, there are two basic forms of Australopithecans: the so-called gracile form (without sagittal crete) and the Paranthropes (South African Paranthropes, Tanzanian Zinjanthropes, Ethiopian mandibles), all with sagittal crete.

Australopithecans still possess a number of simian characteristics, in particular a small brain volume of between 400 and 600 cm^3 , and a prognathic face. However, all these characteristics were less marked than in Anthropoides.

In other words, Australopithecans were human: the occipital orifice was in the vertical plane, the teeth were humanoid, and they were able to stand upright.

What's more, the Australopithecans already had a crude lithic industry. They also used bones to make tools.

Four species of Australopithecus are known:

- *Australopithecus afarensis* or Pre-australopithecus

Known since: 5 MA, most recent: 2.7 MA
Size: 1.10 to 1.30 m
Cranial volume: 300 to 400 cm3
Bipede
Location: Africa

- *Australopithecus africanus*

Known since 3 MA, most recent: 2.2 MA
Height: 1.30 m
Cranial volume: 400 to 600 cm^3
Bipede

Location: Africa

- Two other species: *Australopithecus boisei* and *Australopithecus robustus.*

In the same deposits as the Australopithecus, fossils of an ancient human type have been discovered, associated with pebble-cut tools. These fossils probably represent the oldest form of the Homo genus, dating back 2 to 1.4 million years. The presence of these early tools earned this type of hominid the name Homo *habilis*.

Characteristics of skilled man or *Homo habilis* :

- Known since: 2MA, most recent: 1.5 MA
- Height: 1.20 m
- Cranial volume: 600 to 770 cm^3
- First tools: fitted rollers
- Location: East Africa

All researchers agree that the *Australopithecans* belong in the human line. Along with *Homo habilis*, they represent the first stage of hominisation.

This very distant phase is still relatively poorly understood and is the subject of much debate. On the other hand, we know more about the hominids who lived between 1 and 1.3 million years ago.

For the sake of simplicity, we distinguish 3 successive phases in the formation of modern man: the Pithecanthropians, the Neanderthals and the *Homo sapiens*.

B. Pithecanthrope or *Homo erectus*

These pithecanthropians include the Pithecanthropes of Java, the Sinanthropes of Pekin and Shensis (China), the Atlanthrope of Ternifere near Oran, the Rabat and Casablanca Men, the Telanthropes of South Africa, certain Oldoway Men (Tanzania) and the Chadanthropes.

They are characterised by the following elements:

- Known since 1.6 million years ago, most recent: 150,000 years ago
- Increase in cranial volume of around 800 to 1200 cm^3
- Bifacial stone tools and,
- The conquest of fire.
- Location: Africa, Europe and Asia (China).

Pithecanthropes can be distinguished from modern humans mainly by their more powerful mandibles and smaller cranial volume.

C. Neanderthals or *Homo sapiens Neanderthalensis*

Neanderthals represent a much more recent evolutionary stage than pithecanthropians. They appeared in Eurasia during the last WURN glaciation, around 70,000 years ago.

He was no taller than 1.55 m, with a much larger cranium than *Homo sapiens*: around 1600 cm^3 . Flint tools were perfected and diversified. Some skeletons were buried intentionally, the first known evidence of sepulture.

D. Present-day man or *Homo sapiens sapiens*

This is the systematic position of man today:

Regne :	Animalia
Branch :	Chordata
Class :	Mammalia
Order :	Primates
Family :	Hominidae
Genre :	Homo
Species :	*Homo sapiens sapiens*

It is known as modern man and existed 100,000 years ago. Location, worldwide. His skull (1300 to 1500 cm^3) is different from that of the Cro-Magnon type (1600 to 1700 cm).[3]

Tools had progressed a long way, but sedentary life only began with agriculture in the Neolithic period.

In conclusion, all these human forms fit together so perfectly that the authors do not agree on the nature of the Primates of the human line.

These stages of human evolution do not correspond to a single, simple line; rather, Australopithecus, Pithecanthropes and Neanderthals form a thicket of branches tending towards three structural stages.

Figure 6.5 *The stages of hominisation*

Hominisation took place in several parts of the world, producing several different branches, only one of which has persisted, evolving into present-day man, because, for biologists, all the present-day human species comes from one line, from a single common ancestor: this is monogenism, which is in opposition to polygenism.

6.5 THE MAJOR INDUSTRIES OF GEOLOGICAL TIME

Parallel to biological evolution, hominisation is characterised by the

creation and use of tools. The tools discovered at prehistoric sites reveal the use of increasingly sophisticated technologies.
Some apes use stones and sticks to defend themselves or to obtain food. But only man knows how to systematically make different types of tool for a specific purpose, and then to perfect his technology.
The tools used by prehistoric man were used to kill and butcher animals, to cut and scrape skins, to cut plants and so on.
The raw material used to make these tools is essentially flint.
The Stone Age is divided into three periods: the Paleolithic (ancient stone), the Mesolithic (intermediate period) and the Neolithic (recent stone, corresponding to the polished stone technique).
The table below shows the main stages in the use of paleolithic stone.
The Neolithic represents a decisive turning point in the history of mankind. It saw the appearance of polished stone tools, the first use of metals, and the beginnings of animal husbandry and agriculture.

Table 6.3 *The main stages in the use of paleolithic stone*

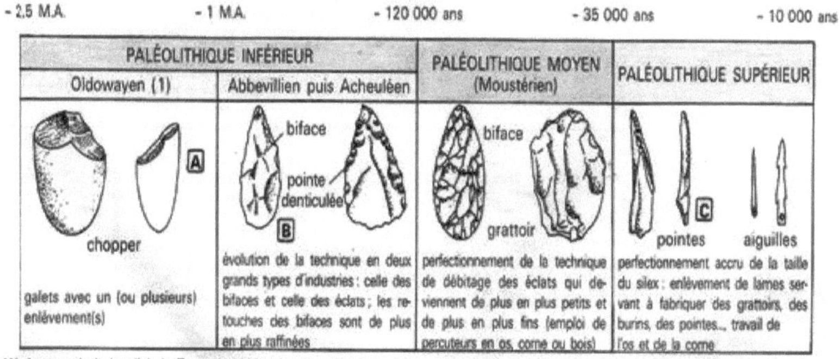

(1) du nom de la localité de Tanzanie (Afrique) qui en fournit les plus nombreux exemplaires (période encore appelée "pebble culture")

In short, tools became increasingly sophisticated over time. Initially made from stone (dressed pebbles, bifaces and flint per^oir in the Paleolithic period), they were later elaborated from a variety of materials (bone, wood and, more recently, metals).

REVISION QUESTIONS ON CHAPTER 6

1. Define these concepts: paleontology, stratigraphy, fossil, microfossil, facies fossil, geological time scale.
2. What is the current subdivision of geological time?
3. Describe the characteristics of the Bonobo.
4. Name a few characteristics that distinguish humans from other primates.
5. What are the major stages of hominisation? Give the characteristics of each of the major stages of hominisation.
6. Describe the main stages in the use of Paleolithic stone, and show how the technique has evolved over time.

BIBLIOGRAPHICAL REFERENCES

Boden J.-P. al., *1988. Biologie Geologie.* Premiere S Collection Tavernier. Paris, Bordas, 352p.

Boden J.-P. al., *1993. Sciences de la Vie et de la Terre*. Premiere S Collection Tavernier-Lizeaux. Paris, Bordas, 400p.

BOILLOT G., HUCHON P. and LAGABRILLE Y. 2020. *Introduction a la geologie, dynamique de la terre.* 5eme edition. Paris, Dunod, 252p.

DAJOZ R., 1970. *Precis d'ecologie*. Paris, Dunod, 357p.

DUVIGNAU P., 1974. *La synthese ecologique*. Paris, Doin, 296p.

FREDERICK K. LUTEGENS and EDWARD J. TARBUCH, 2016. *Essentials of geology.* 13th edition. New York, Person, 609p.

JOHN P. RAFFERT, 2012. *Geologicalsciences*. New York, Britannica, 226p.

KATABANA J.-P., 1991. *Geologie de l'ingenieur*. Notes de cours. Inedites. Universite de Butare, 175p.

MICHAEL L. CAN, HANS DRAMMAN, ROBERT A. LUE and CAROL KAESUK YOON, 2006. *Discovering biology*. 2nd edition. Bruxelles, Editions De Boeck Universite, 727p., glossary and index. A French adaptation by Ana PERSIC, Helene ROCHER and Rachel VINCENT.

NEIL A. CAMPBELL and JANE B. REECE, 2007. *Biology*. 3rd edition. Quebec, Editions du Renouveau Pedagogique, 1334 pages plus appendices. A French adaptation by Rene Lechaine and Michel Bosset.

PEYCRU P. et al. 2008. *Geologie*. Paris, Dunod, 658p.

POTY E., 2013. *Geologie*. Departement de l'Universite de Liege, 150p.

SOREL D. and VERGELY P., 2018. *Atlas d'initiation aux cartes et coupes geologiques*. 4eme edition. Paris, Dunod, 128p.

NEIL A. CAMPBELL and JANE B. REECE, 2007. *Biology.* 3rd edition. Quebec, Editions du Renouveau Pedagogique, 1334 pages plus appendices. A French adaptation by Rene Lechaine and Michel Bosset.

MICHAEL L. CAN, HANS DRAMMAN, ROBERT A. LUE and CAROL KAESUK YOON, 2006. *Discovering biology*. 2nd edition. Bruxelles, Editions De Boeck Universite, 727p., glossary and index. A French adaptation by Ana PERSIC, Helene ROCHER and Rachel VINCENT.

SOME WEBSITES CONSULTED

Philippe DUCHAUFOUR, "HUMUS", Encyclopedia Universalis [online], URL : https://www.universalis.fr/encyclopedie/humus/ consulted on 4 December 2022.

URL: https://photo.geo.fr/15-volcans-parmi-les-plus-dangereux-au-monde- 50964#le-krakatoa-indonesie-d1que Accessed 30 November 2022.

APPENDICES

ASTRONOMICAL DATA

(Source IMCCE)

Periods of the moon's sideral revolution around the earth and of its own rotation: 27d 7h 43min = 27.321d.

Mass of the Moon: $M_L = 0.074 \times 10^{24}$ kg = 0.012 M_T

Average lunar radius: 1,740km.

Average distance Earth-Moon: 384,400 km

Mass of sun =1.9889×10^{30} = 333,948M_T.

Radius of the sun =0.696×10^{6} km.

Earth-Sun distance = 149.6×10^{6} km = 1 A.U.

Distance from the sun to the asteroid belt = 2-3 A.U.

AUTHOR'S CONTACT DETAILS

Albert ALOMBA A. Biology and chemistry teacher at the Groupe Scolaire Consulaire Congolais in Kigali.
Telephones: +250 788 540 257 or +250 738 540 257
E-mail addresses: alomba2@gmail.com or alomba2@yahoo.fr

MIX
Papier aus verantwortungsvollen Quellen
Paper from responsible sources
FSC® C105338

Printed by Books on Demand GmbH, Norderstedt / Germany